Windows® 2000 Registry

2ND EDITION

Little Black Book

Nathan Wallace
Anthony Sequeira

Publisher
Steve Sayre

Acquisitions Editor
Charlotte Carpentier

Development Editor
Jessica Choi

Product Marketing Manager
Tracy Rooney

Project Editor
Greg Balas

Technical Reviewer
Jim Kelly

Production Coordinator
Peggy Cantrell

Cover Designer
Jody Winkler

Layout Designer
April E. Nielsen

Windows® 2000 Registry Little Black Book

Limits of Liability and Disclaimer of Warranty

The author and publisher of this book have used their best efforts in preparing the book and the programs contained in it. These efforts include the development, research, and testing of the theories and programs to determine their effectiveness. The author and publisher make no warranty of any kind, expressed or implied, with regard to these programs or the documentation contained in this book.

The author and publisher shall not be liable in the event of incidental or consequential damages in connection with, or arising out of, the furnishing, performance, or use of the programs, associated instructions, and/or claims of productivity gains.

Trademarks

Trademarked names appear throughout this book. Rather than list the names and entities that own the trademarks or insert a trademark symbol with each mention of the trademarked name, the publisher states that it is using the names for editorial purposes only and to the benefit of the trademark owner, with no intention of infringing upon that trademark.

The Coriolis Group, LLC
14455 North Hayden Road
Suite 220
Scottsdale, Arizona 85260

(480) 483-0192
FAX (480) 483-0193
www.coriolis.com

Library of Congress Cataloging-in-Publication Data
Wallace, Nathan.
    Windows 2000 registry little black book / by Nathan Wallace and Anthony Sequeira.
    –2nd ed.
        p. cm.
        ISBN 1-57610-882-1
        1. Microsoft Windows (Computer file)   2. Operating systems (Computers)
I. Sequeira, Anthony.   II. Title.
QA76.76.063 W3584 2001
005.4'4769–dc21   2001028457
                                                                                            CIP

Printed in the United States of America
10   9   8   7   6   5   4   3   2   1

The Coriolis Group, LLC • 14455 North Hayden Road, Suite 220 • Scottsdale, Arizona 85260

A Note from Coriolis

Coriolis Technology Press was founded to create a very elite group of books: the ones you keep closest to your machine. In the real world, you have to choose the books you rely on every day *very* carefully, and we understand that.

To win a place for our books on that coveted shelf beside your PC, we guarantee several important qualities in every book we publish. These qualities are:

- *Technical accuracy*—It's no good if it doesn't work. Every Coriolis Technology Press book is reviewed by technical experts in the topic field, and is sent through several editing and proofreading passes in order to create the piece of work you now hold in your hands.

- *Innovative editorial design*—We've put years of research and refinement into the ways we present information in our books. Our books' editorial approach is uniquely designed to reflect the way people learn new technologies and search for solutions to technology problems.

- *Practical focus*—We put only pertinent information into our books and avoid any fluff. Every fact included between these two covers must serve the mission of the book as a whole.

- *Accessibility*—The information in a book is worthless unless you can find it quickly when you need it. We put a lot of effort into our indexes, and heavily cross-reference our chapters, to make it easy for you to move right to the information you need.

Here at The Coriolis Group we have been publishing and packaging books, technical journals, and training materials since 1989. We have put a lot of thought into our books; please write to us at **ctp@coriolis.com** and let us know what you think. We hope that you're happy with the book in your hands, and that in the future, when you reach for software development and networking information, you'll turn to one of our books first.

Coriolis Technology Press
The Coriolis Group
14455 N. Hayden Road, Suite 220
Scottsdale, Arizona
85260

Email: ctp@coriolis.com
Phone: (480) 483-0192
Toll free: (800) 410-0192

Look for these related books from The Coriolis Group:

Windows 2000 Active Directory Black Book
By Adam Wood

Windows 2000 Security Little Black Book
By Ian McLean

Windows 2000 Server Architecture and Planning, 2nd Edition
By Morten Strunge Nielson

Windows 2000 System Administrator's Black Book, 2nd Edition
By Stu Sjouwerman, Barry Shilmover, and James Michael Stewart

Windows Admin Scripting Little Black Book
By Jesse M. Torres

Also published by Coriolis Technology Press:

SQL Server 2000 Black Book
By Patrick Dalton and Paul Whitehead

Windows 2000 Professional Upgrade Little Black Book
By Nathan Wallace

Windows 2000 Professional Advanced Configuration and Implementation
By Morten Strunge Nielsen

Windows 2000 Systems Programming
By Al Williams

Windows 2000 TCP/IP Black Book
By Ian McLean

To Roger Aldrich. We miss you.
—Nathan Wallace

❧

This book is dedicated to my Mom, Miss Donna Lee. She not only
gave me life, but also taught me how to live it to the fullest.
—Anthony Sequeira

❧

About the Authors

Nathan Wallace (Lakewood, CO) has been a Windows power user since version 1, and has been a best-selling computer book author since 1995, with 16 books in print and almost 100,000 copies sold. He is a Microsoft Sitebuilder Network Level 2 member specializing in Active Desktop integration and ActiveX scripting.

Anthony Sequeira (Fountain Hills, AZ) has been a professional speaker and writer in the IT industry for the last seven years. He holds every major Microsoft certification including MCT, MCSE+I, MCSE 2000, MCDBA, and MCSD. Anthony's books include Windows 2000 Server and Active Directory titles. He currently speaks about Active Directory technologies and Cisco networking for KnowledgeNet.com.

Acknowledgments

First and foremost, we'd like to thank Charlotte Carpentier, Acquisitions Editor at Coriolis. Also, special thanks to Greg Balas, who served as Project Editor, and Peggy Cantrell, who served as the Production Coordinator for the book. In addition, we'd like to thank Tracy Rooney, our Marketing Specialist; April Nielsen, for creating the book's interior design; and Laura Wellander, for designing the cover. Each of these folks made significant contributions to the final book, and made it much better than the average technical tome. We would also like to thank Louise Kohl Leahy for copyediting the book, Jim Kelly for performing the technical review, Christine Sherk for proofreading, and Janet Perlman for indexing this Second Edition. Finally, thanks to all the Microsoft programmers who made Windows 2000 possible. Despite a national sport of Microsoft-bashing, we still get the best operating systems around from Redmond.
—*Nathan Wallace*
—*Anthony Sequeira*

Contents at a Glance

Table of Contents

Chapter 5
Windows 2000 User Interfaces 85

Chapter 8
General Networking 153

Chapter 11
Routing and Remote Access Service 205

Chapter 19
Microsoft Transaction Server 335

Introduction

The specific goal of this book is to save Windows 2000 administrators hours and hours of fruitless searching for solutions to perplexing and potentially disastrous problems. When something goes wrong with a Windows 2000 application or service, many times the solution lies in a Registry entry. This book is intended to be a step-by-step guide for Windows 2000 administrators to help them locate and correct Registry-related problems.

Who This Book Is For

This book is not for amateurs; it requires a willingness to directly change Registry entries using Regedt32.exe, a process that can in theory significantly impact Windows 2000. Administrators must fully understand how to back up the Registry and restore it before starting any of the changes suggested in this book.

Once an administrator is familiar with using Regedt32.exe and backing up the Registry, this book holds a wealth of solutions to critical Windows 2000 failures and bottlenecks. From problems with Internet Explorer and TCP/IP networking; through SQL and Office; to IIS, Windows 2000 shells, and local and networked printers, this book has hundreds of immediate solutions to problems that otherwise might require days of effort to correct. This book draws from dozens of resources from the Internet, Microsoft's Knowledge Base, and the many books and magazine articles on Window 2000 and the Registry. What you hold in your hands is a distillation of the work and wisdom of hundreds of Windows 2000 professionals. Just take this book back to your desk and it will pay for itself a dozen times

What You Should Know

Almost every Immediate Solution presented in this text deals with a Registry modification using Regedt32. Many of the changes take place in **HKEY_LOCAL_MACHINE**, and most of these changes require a reboot of the Windows 2000 system to take effect. Other changes occur in **HKEY_CURRENT_USER**, and these changes simply require a log off and then log on in order for the changes to take effect.

You should also realize that some of the Registry changes presented in the Immediate Solutions can also be implemented using a Graphical User Interface somewhere in Windows 2000. They are still included in this text because it is often much more efficient to make the change using the Registry, especially when you need to change the setting for a large number of systems.

What's Inside

The book is divided into 21 chapters. Each chapter contains a brief overview of the topic, followed by a set of practical immediate solutions to show you how to effectively complete tasks and solve problems.

Chapter 1: Overview of the Registry

Chapter 1 covers the layout and format of the Windows 2000 Registry, including the 11 recognized Registry data types and the hives that contain all the data in the Registry. It provides complete coverage on using Regedt32.exe to locate and manipulate all the types of Registry data available in Windows 2000.

Chapter 2: Registry Tools

Chapter 2 covers the various ways to manipulate the Registry, including Computer Management, Regedt32, Control Panel applets, and the Windows 2000 Resource Kit tools.

Chapter 3: Registry Disaster Prevention

Chapter 3 explores the various ways to safeguard the Registry. This is particularly important when you are making frequent modifications to this important database.

Chapter 4: System Administration Tools

Chapter 4 covers the Performance Monitor and several other administration tools, including ways to prevent counter timeouts, control log creation and format, deal with UNICODE, and manipulate paged and nonpaged memory pool sizes.

Chapter 5: Windows 2000 User Interfaces

Chapter 5 covers the three Windows 2000 shells: command prompt, Program Manager, and Explorer. Immediate solutions include file name auto completions, command window positioning, Explorer automatic restarts, suppression of Explorer window reopening on restart, adding file extensions to the New shortcut menu, and setting individual user preferences for logon screen color, wallpaper, and keyboard settings.

Chapter 6: TCP/IP and the Internet

Chapter 6 covers TCP/IP networking with Windows 2000, an important subject because almost all Windows 2000 networks use TCP/IP. Immediate solutions include dead gateway detection, increasing buffer memory for routing, setting NetBEUI keep-alive and node type values, and setting the threads allocated to WINS.

Chapter 7: Hardware and Systems

Chapter 7 covers HAL, Services, and WOW. Immediate solutions include switching to the LKG (Last Known Good) Registry copy on service startup failure, controlling Windows 2000 service startup order, activating SCSI driver debugging breakpoints, and preventing WOW device timeouts.

Chapter 8: General Networking

Chapter 8 covers general networking, including browser, UPS, and workstation services, as well as directory replication. Immediate solutions include enhancing Named Pipe performance, setting opportunistic locking and raw reading configurations for improved I/O, preventing lost mailslot messages, enabling Novell drives to work as Windows 2000 drives, and configuring the directory replication service.

Chapter 9: Networking Protocols and Interoperability

Chapter 9 covers networking protocols such as AppleTalk, Novell, and NetBIOS. Immediate solutions include disabling SAP dialouts, controlling SPX timeouts, fixing MacFile Server volume path errors, setting the AppleTalk router, and updating NDIS bindings on a configuration change.

Chapter 10: Local and Networked Printers

Chapter 10 covers local and networked printers. Immediate solutions include slowing down fast printing, setting a printer's priority class, setting a printer's spooler priority, setting a networked printer's name, and enabling trusted printing.

Chapter 11: Routing and Remote Access Service

Chapter 11 covers Remote Access Service (RAS), including protocol support, auditing, TCP/IP, and PPP. Immediate solutions cover enabling RAS auditing and port logging, promoting RAS NetBIOS traffic over multicast traffic, enabling simultaneous network connections to improve RAS performance, forcing RAS serial FIFO queuing, and changing the RAS WINS server name.

Chapter 12: Windows 2000 Security

Chapter 12 covers Windows 2000 Security, including access tokens, SAM Registry keys, logging on, and Netlogon. Immediate solutions cover setting the lifetime of Security Event log entries, forcing a Windows 2000 crash when the Security event log is full, preventing BDC pulse timeouts and traffic jams, and establishing C2 security for CD-ROM and floppy drives.

Chapter 13: Windows 2000 Help

Chapter 13 covers the powerful new Windows 2000 Help system, with details on the Registry settings for CHM and CHI files, information on dealing with potent Troubleshooter technology, and how to fix a broken Visual Studio MSDN installation.

Chapter 14: System Multimedia

Chapter 14 covers System Multimedia, a facet of Windows 2000 brought over from Windows 9x. Information is provided on configuring and troubleshooting the video MediaPlayer, the powerful and multifaceted DirectX technology, and the older MCI (Media Control Interface) settings and user-installable sound drivers.

Chapter 15: Registry Programming

Chapter 15 describes methods for programming with the Registry in mind. Microsoft provides tools to simplify this endeavor and this chapter allows you to quickly focus on this very important aspect of Windows programming.

Chapter 16: COM+

Chapter 16 covers COM+, the new incarnation of Component Object Model technology that is a core aspect of Windows 2000. Elements include adding remote COM+ computers, enabling COM+ router functionality, and configuring the amazing new IMDB (In-Memory Database) technology for maximum performance.

Chapter 17: Internet Information Server

Chapter 17 covers Internet Information Server (IIS), including the FTP and HTTP/WWW services, as well as general settings. Immediate solutions include setting the IIS thread pool for I/O wait background processing, adjusting IIS TTL (time to live) cache settings, preventing IIS slow connection timeouts, disabling guest logons for FTP and WWW services, and controlling IIS log file creation and settings.

Chapter 18: Internet Explorer 4 +

Chapter 18 covers Internet Explorer and the Active Desktop for Windows 2000. It includes immediate solutions for controlling URL display in IE, controlling access to the script debugger and display of menus for the Active Desktop, and setting IE safety warnings and colors on a per-user basis.

Chapter 19: Microsoft Transaction Server

Chapter 19 covers Microsoft Transaction Server (MTS). Its immediate solutions show how to determine the installation paths for MTS, as well as how to determine whether packages are installed and if so, what components they contain, which Security Roles are enabled and which users are in them, and how the threading and transaction settings are made for MTS components.

Chapter 20: SQL Server

Chapter 20 covers SQL Server, with special emphasis on SQL Executive and DB-library interactions. Immediate solutions include configuring Named Pipes, fixing Executive logon problems, and preventing startup failures due to slow RPC initialization.

Chapter 21: Microsoft Office

Chapter 21 explores key Registry values associated with the most popular office suite of applications in use today.

Overview of the Registry

In Brief

The Windows 2000 Registry is a binary database holding most of the information the operating system needs to function. Each time a Windows 2000 computer boots, the Registry is reconstructed from a set of files that are updated when the computer shuts down. Once in memory, the Registry is maintained continually. The Registry functions as a relational database of system parameters and is in many ways the clearinghouse for a Windows 2000 computer's behavior and capabilities. There are two basic Registry areas that you must understand before you can uncover its deeper secrets—its hierarchical structure of hive keys, keys, subkeys, and values, and its various Registry data types.

The Windows 2000 Registry Structure

The information in the Windows 2000 Registry is in binary format rather than the text format of Windows 3.x INI files or MS-DOS system files because the Registry is so large and must be accessed quickly to avoid degrading system performance. To minimize the cost in time of obtaining a given bit of information from this huge store of data, it is organized like a file system on the hard drive.

The Windows 2000 Registry is organized into four levels, in a descending hierarchy:

- *Hive keys*—There are five system-defined hive keys, known by the first portion of their name: **HKEY_**.

- *Keys*—There are user-defined and system-defined keys. These keys have no special naming convention and they exist as subdirectories of the main **HKEY_** hive keys. Most keys and subkeys have no associated data; they serve only to organize access to data.

- *Subkeys*—There are user-defined and system-defined subkeys. These have no special naming convention and exist as subdirectories of the user-defined or system-defined keys. Most keys and subkeys have no associated data; they serve only to organize access to data. (Note that some documentation makes no distinction between keys and subkeys.)

- *Value entries*—These elements are at the bottom of the hierarchy and are like files in a file system. They contain actual data that are used in the performance of the computer and its applications. Values are classified by a small, but effective, set of available data types, which are covered later in this chapter.

Now, let's examine each of the hive keys in turn, looking at what information they contain and how they interact with the Windows 2000 system.

HKEY_LOCAL_MACHINE

HKEY_LOCAL_MACHINE (HKLM) is the hive key that contains operating system and hardware-oriented information, such as the bus type of the computer, total system memory available, currently loaded device drivers, and the bootup control data. In fact, HKLM holds most of the information in the Registry, because two of the other four hive keys are aliased to its subkeys. Figure 1.1 shows the HKLM hive in Regedt32 under Windows 2000. The other hive keys have similar appearances and organization.

TIP: *Hive keys are usually abbreviated "HK" plus the first initials of the component words in the underscored space.*

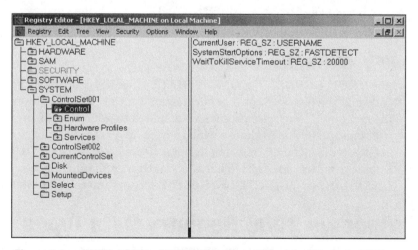

Figure 1.1   **HKEY_LOCAL_MACHINE** in Regedt32 under Windows 2000.

HKEY_CURRENT_USER

The *HKEY_CURRENT_USER (HKCU)* hive key contains the user profile for the user currently logged in on the computer the Registry serves. Its subkeys include environment variables, personal program groups, desktop settings, network connections, printers, and application preferences. (Windows 2000 uses *Environmental variables* to allow scripts, Registry entries, and other applications to use wildcards for important system information that might change.) This information is mapped from the **HKEY_USERS** hive key's Security ID (SID) subkey for the currently logged on user.

HKEY_CLASSES_ROOT

The *HKEY_CLASSES_ROOT (HKCR)* hive key contains subkeys listing all the COM servers currently registered on the computer and all the file extensions currently associated with applications. This information is mapped from the **HKEY_LOCAL_MACHINE\SOFTWARE\Classes** subkeys.

HKEY_USERS

The *HKEY_USERS (HKU)* hive key contains subkeys that contain all the user profiles for the current computer. One of its subkeys is always mapped to **HKEY_CURRENT_USER** (via the user's SID value). Another subkey, **HKEY_USERS\DEFAULT**, controls the Windows settings that take effect before a user presses Ctrl+Alt+Del to log on.

HKEY_CURRENT_CONFIG

The *HKEY_CURRENT_CONFIG (HKCC)* hive key contains subkeys listing all the hardware profile information for the current session of the computer. Hardware profiles were added to Windows NT in version 4 and allow you to choose which device drivers to load for a given session on a machine (for example, not activating a docking port when a laptop is not docked). This information is mapped from the **HKEY_LOCAL_MACHINE\SYSTEM\CurrentControlSet** subkeys.

Windows 2000 Registry Data Types

Because the Registry is a small relational database, it logically needs a schema to define its organization. This structure, aside from the hierarchy outlined above, is provided by a restricted set of data types that Registry values can contain. There are 11 recognized data types for Windows 2000 Registry entries. Table 1.1 provides a list of their names and capabilities.

Table 1.1   Recognized data types for the Windows 2000 Registry.

Data Type	Raw Format Type	Function
REG_NONE	Unknown	Encrypted data
REG_SZ	String	Text characters
REG_EXPAND_SZ	String	Text with variables
REG_BINARY	Binary	Binary data
REG_DWORD	Number	Numerical data
REG_DWORD_BIG_ENDIAN	Number	Non-Intel numbers
REG_LINK	String	Path to a file
REG_MULTI_SZ	Multi-string	String arrays
REG_RESOURCE_LIST	String	Hardware resource list
REG_FULL_RESOURCE_DESCRIPTOR	String	Hardware resource ID
REG_RESOURCE_REQUIREMENTS_LIST	String	Hardware resource ID

Most of these raw format types has a dedicated editor in Regedt32, namely Number, Binary, String, and Multi-String. These editors are covered in the "Immediate Solutions" section.

Immediate Solutions

Accessing the Registry with Regedt32

Windows 2000 provides an excellent Registry editing utility, *Regedt32*. The easiest way to get to the Registry Editor is to do the following:

1. Open the Start menu and select Run.

2. Browse by clicking on the Browse button until you locate your Windows 2000 root directory (usually WINNT).

3. Move down to the System32 directory and locate Regedt32.exe. Select it into the dialog box and click on the OK button. (Some Windows 2000 documentation mentions another Registry Editor, Regedit.exe. However, that editor lacks many of the useful features of Regedt32.exe, so we won't cover it for this book.) See Figure 1.2.

The Registry Editor application begins with all its hives in separate cascaded windows.

TIP: *To avoid making accidental changes to the Registry, you can choose Options/Read Only Mode. The menu option is a toggle, so choosing it again restores the Registry to editable mode. There will be more on this later in the chapter.*

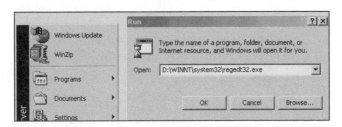

Figure 1.2   Starting the Regedt32 Registry Editor from the Start menu.

Accessing the Registry of a Remote System

You can use Regedt32 to access and manipulate the Registry on remote Windows systems. Obviously, you must have the appropriate permissions on the remote system in order to do this. To access the Registry on a remote system:

1. From the Registry menu, click on Select Computer.

2. In the Computer field type the name of the computer whose Registry you want to open and click on OK. (Or you can select the computer from the Select Computer area.)

3. When you are finished with the remote Registry, you can choose Open Local from the Registry menu.

Searching for a Key

Regedt32 offers a very basic Find utility that can help in a quest for a configuration change. To use it, follow these steps:

1. Launch Regedt32.

2. Click on the View menu and choose Find Key.

3. In the Find dialog box, complete the Find What: field with the name of the key for which you are looking and click on the Find Next button. Note the options for matching case and word completion.

Adding New Registry Keys and Values

Microsoft warns continually about the danger of modifying the Registry manually. There is good reason for this. Damaging the Registry in a way not immediately detected by the system can cause the damaged copy to be considered *last known good* (the backup copy maintained for problems with Registry corruption), which destroys the information needed to restore the system by replacing the previous (working) copy. Chapter 3 covers all the steps you can take to ensure that disaster does not strike due to modifications to the Registry.

Here is how to manually add a Registry key and value:

1. Launch Regedt32.

2. Select Window|**HKEY_LOCAL_MACHINE**.

3. Use the tree control in the left-hand window to navigate to the Software subkey. Double-click on the subkey to expand it.

4. Select the Edit|Add Key menu option. The Add Key dialog box appears, as shown in Figure 1.3.

5. Enter the name of the dummy key in the Key Name field. Make sure it is different from any of the other subkeys on its level and that the name does not contain a backslash (\) character. In the Class edit control, enter "REG_SZ" to indicate a string value. Click on OK.

6. Select the newly added subkey in the tree control in the left-hand window.

7. Choose the Edit|Add Value menu option. The Add Value dialog box appears, as shown in Figure 1.4.

8. Enter the name of the value item in the Value Name edit control. From the Data Type dropdown list, choose REG_SZ as the data type. Click on OK.

9. The String Editor dialog box appears, allowing you to enter the data you want to save. Click on OK to save the new value and data.

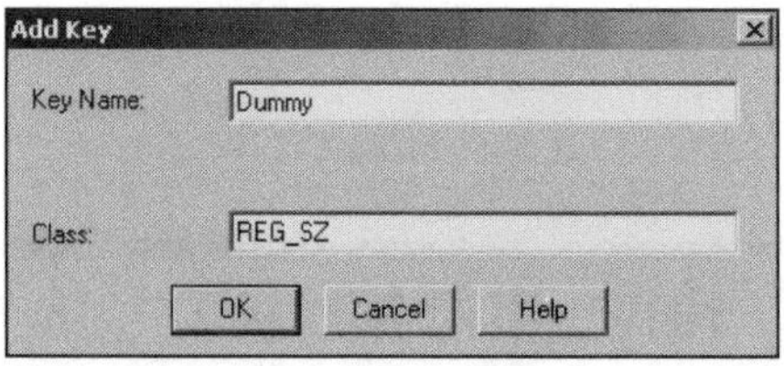

Figure 1.3   The Registry Editor's Add Key dialog box.

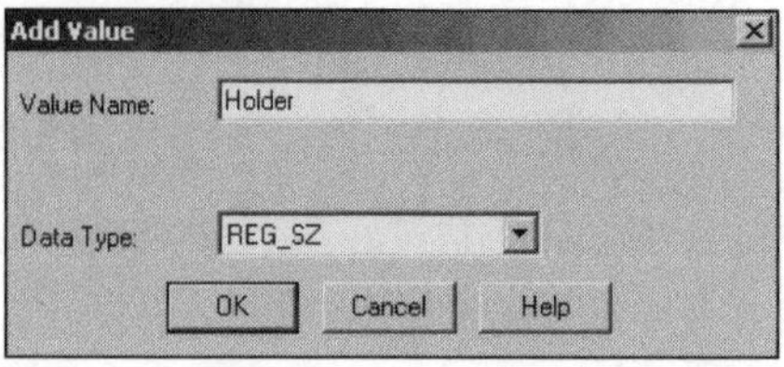

Figure 1.4   The Registry Editor's Add Value dialog box.

Modifying or Deleting an Existing Registry Key

You will want to modify the data in an existing Registry key more often than you'll want to add a new key. Also, you might sometimes need to delete a Registry key. To perform either of these tasks, follow these steps:

1. Launch Regedt32.

2. Select the Window menu option that matches the hive key under which you want to work.

3. Use the tree control in the left-hand window to navigate to the desired subkey. Expand the subkey by double-clicking on it.

4. Click on the value entry you want to change.

5. On the Edit menu, select the appropriate Edit command for the data type of the value (for example, choose String for the REG_SZ type). A dialog box appears allowing you to change the value entry. Click on OK to make the change.

TIP: *You can trigger an automatic selection in the editing dialog box by simply double-clicking on the value in the right-hand window.*

---

6. To delete either a value or a subkey (you cannot delete hive keys), select Delete.

7. A Warning dialog box appears and you must click on OK to continue the deletion. If you do so and the selected item is a key or subkey, all related subkeys and value entries are deleted. If the selection is a value entry, only it and its information will be deleted.

WARNING! *There is no Undo option! Use deletion with extreme care.*

Related solution:	Found on page:
Saving Registry Keys with Registry Editor	66

Modifying a Registry Value of Binary Data Type

Some Registry value entries are raw binary data. Interestingly enough, you can modify these entries from the Registry Editor. Follow these steps:

1. Launch Regedt32.

2. Select the Window menu option that matches the hive key under which you want to work.

3. Use the tree control in the left-hand window to navigate to the desired subkey. Expand the subkey by double-clicking on it.

4. Click on the Binary value entry you want to change.

5. On the Edit menu, select Binary. The Binary Editor appears, as shown in Figure 1.5.

6. As you can see in Figure 1.5, you have two options with this dialog box: You can set it for pure Binary (1 and 0) or you can set it for Hexadecimal (digits 0 through 9 and the letters A through F). Select the radio button of whichever mode is easier for your operation.

7. You can either overwrite the existing information or add to it. The editor supports normal edit control selection operations. The numbers along the top give the byte count of the data stream. When you are done, click OK to add the updated value to the Registry.

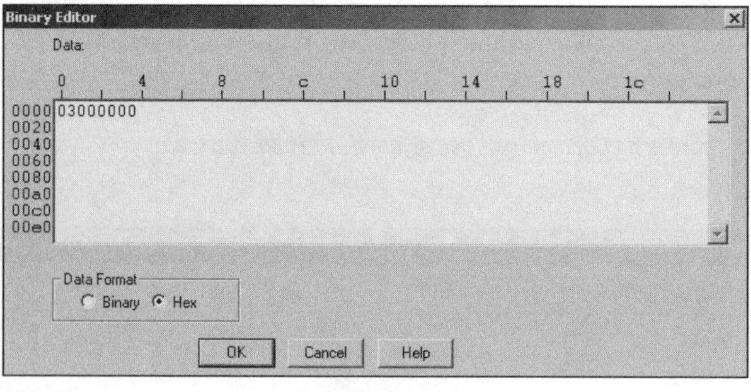

Figure 1.5   The Registry Editor's Binary Editor dialog box.

Modifying a Registry Value of String Data Type

Most Registry data is in string format, type REG_SZ, or one of the similar data types. To edit string data in the Registry with the Registry Editor, follow these steps:

1. Launch Regedt32.

2. Select the Window menu option that matches the hive key under which you want to work.

3. Use the tree control in the left-hand window to navigate to the desired subkey. Expand the subkey by double-clicking on it.

4. Click on the value entry you want to change.

5. On the Edit menu, select String. The String Editor appears, as shown in Figure 1.6.

6. As you can see in Figure 1.6, the String Editor is simply an edit control. It allows you to overwrite the existing information or add to it. The editor supports normal edit control selection operations. When you are finished, click on OK to add the changed value to the Registry.

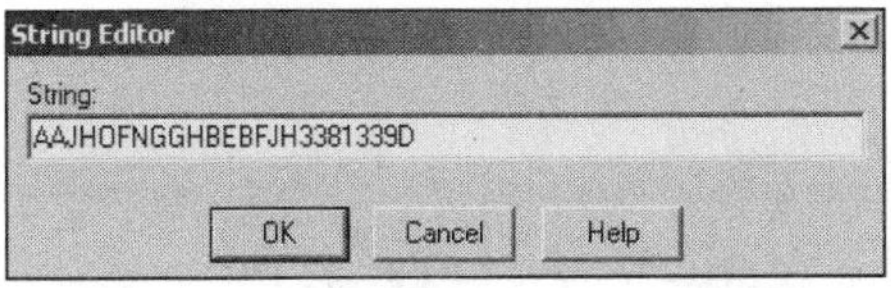

Figure 1.6   The Registry Editor's String Editor dialog box.

Modifying a Registry Value of Expandable String Data Type

A more sophisticated type of string-based Registry data type is the *expandable string* (REG_EXPAND_SZ). To edit an expandable string value in the Registry with the Registry Editor, do the following:

1. Launch Regedt32.

2. Select the Window menu option that matches the hive key under which you want to work.

3. Use the tree control in the left-hand window to navigate to the desired subkey. Expand the subkey by double-clicking on it.

4. Click on the value entry you want to change.

5. On the Edit menu, select String. The String Editor appears, as shown in Figure 1.7.

6. As you can see in Figure 1.7, the String Editor is simply an edit control. It allows you to overwrite the existing information or add to it. The editor supports normal edit control selection operations. The expandable portion of the string is included inside the percent characters. You must be careful when you change the expandable portion of the string so that it matches one of the known environment variables or the value will cease to function correctly. When you are finished, click on OK to add the changed value to the Registry.

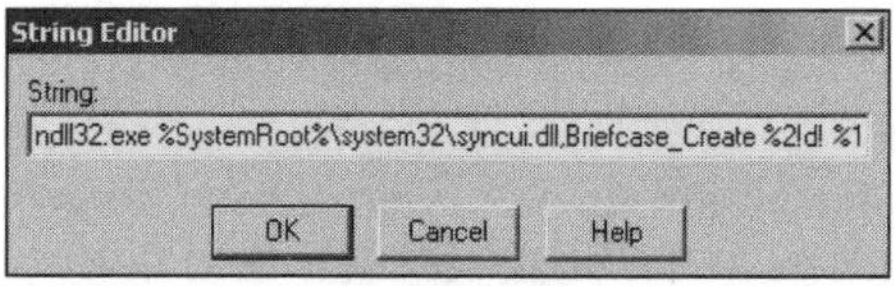

Figure 1.7   The Registry Editor's String Editor dialog box, showing an expandable string.

Modifying a Registry Value of DWORD Data Type

DWORD data types are the only way to store numbers in the Registry. To edit a DWORD value, follow these steps:

1. Launch Regedt32.

2. Select the Window menu option that matches the hive key under which you want to work.

3. Use the tree control in the left-hand window to navigate to the desired subkey. Expand the subkey by double-clicking on it.

4. Click on the value entry you want to change.

5. On the Edit menu, select DWORD. The DWORD Editor appears, as shown in Figure 1.8.

6. As you can see in Figure 1.8, the DWORD Editor is an edit control. It allows you to overwrite the existing information or add to it. The editor supports normal edit control selection

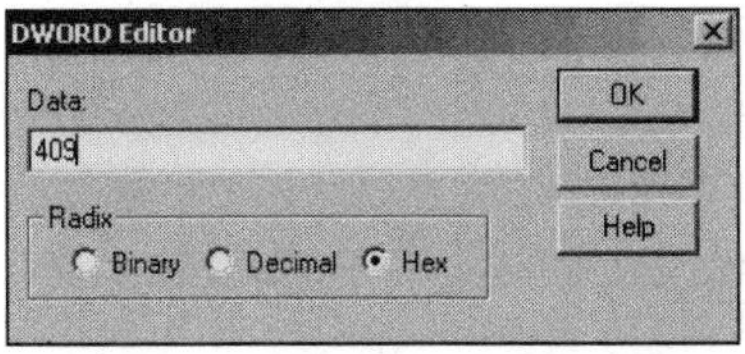

Figure 1.8   The Registry Editor's DWORD Editor dialog box.

operations. The radio buttons at the bottom determine whether the data is in Binary (1 and 0), Decimal (0 through 9), or Hexadecimal (0 through 9 and A through F) format. Select the format that makes the most sense for the value you are editing. When you are done, click on OK to add the changed value to the Registry.

Modifying a Registry Value of Multi-String Data Type

Some Registry value entries are groups of strings. Rather than having to make each such string a separate value, the Registry provides the REG_MULTI_SZ data type to enter and edit string arrays. You accomplish these tasks as follows:

1. Launch Regedt32.

2. Select the Window menu option that matches the hive key under which you want to work.

3. Use the tree control in the left-hand window to navigate to the desired subkey. Expand the subkey by double-clicking on it.

4. Click on the value entry you want to change.

5. On the Edit menu, select Multi-String. The Multi-String Editor appears, as shown in Figure 1.9.

6. As you can see in Figure 1.9, the Multi-String Editor is a multiline edit control. It allows you to overwrite the existing information or add to it. The editor supports normal edit control selection operations. Each separate string has its own line in the editor. When you have finished your changes, click on OK to add the changed value to the Registry.

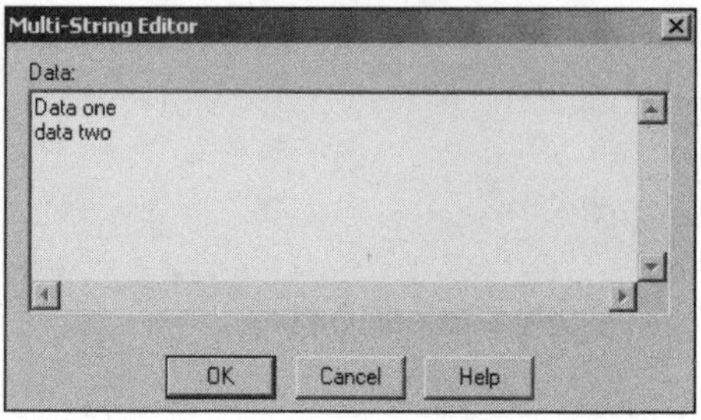

Figure 1.9   The Registry Editor's Multi-String Editor dialog box.

Related solutions:	*Found on page:*
Saving Registry Keys with Registry Editor	66
Restoring Registry Keys with Registry Editor	67

Setting Security on the Registry Using Regedt32

Regedt32 allows you to set unique security settings on any key in the Registry. To change security settings, follow these steps:

1. From within Regedt32, choose the key you want to secure.

2. Select the Security menu.

3. Choose Permissions.

4. In the Permissions window, use the following options to configure security on the key:

   - The Name area displays all the accounts that have levels of permissions on the key. You can select an account to view its specific permissions.

   - The Permissions area displays the permissions for the accounts selected in the Name area.

   - The Advanced button opens the Access Control Settings window, which allows you to set more advanced permissions, as well as auditing and ownership.

   - The Allow Inheritable Permissions From Parent to Propagate To This Object checkbox permits security permissions from parent objects to flow down through inheritance to this key. Note this is the default behavior.

> **TIP:** *To view an example of very restrictive security settings in the Registry, access the Registry of a domain controller and navigate to* **HKLM\SECURITY\SAM**. *Use the Security menu to view the permissions—or lack thereof.*

Regedt32 Options

There are several options available in Regedt32 that can assist you quite dramatically. The options are available via the Options menu or (of course) through Registry settings themselves. The options are as follows:

- *Font*—Allows you to configure the font used in Regedt32 including the font's properties. This is very useful if you have trouble seeing the default font.

- *Auto Refresh*—Automatically updates the Registry when any change is made to Registry data (this is enabled by default).

- *Read Only Mode*—This allows you to completely explore the Registry without worrying about inadvertent changes.

- *Confirm On Delete*—This option causes a dialog box to appear, asking a user to confirm whether he or she wants to delete a key or value in the Registry.

- *Save Settings On Exit*—This option causes Regedt32 to open focused on the hive that had focus when Regedt32 was last closed.

You can also set these options via the Registry:

1. Launch Regedt32.

2. Select the Window menu option for **HKEY_Current_User**.

3. Use the tree control in the left-hand window to navigate to the **SOFTWARE\Microsoft\RegEdt32\Settings** subkey. Double-click on the subkey to expand it. Its values appear in the right-hand window.

4. Notice the REG_SZ values for the Regedt32 options. A value of 1 enables the feature, while a value of 0 disables the feature.

5. Because the Save Settings value resets to 1 by default, be sure to use the Permissions option from the Security menu to deny the user read access to the Registry key once you have made that change.

1. Overview of the Registry

Changing the Maximum Size Allowed for the Registry Database

You can control the total size allowed for the Registry database. Registry information is stored in the paged pool. The paged pool is a portion of virtual memory that is written to physical disk when not in use. A Registry value named **RegistrySizeLimit** allows an administrator to prevent the operating system from completely filling the paged pool with Registry information.

By default, the Registry size limit is 33 percent of the size of the paged pool. A Registry can be up to 80 percent of the size of the paged pool. Thanks to a Registry modification, you can actually change the size limit for the Registry on a Windows 2000 system. You should only consider changing the Registry size if the computer is a domain controller for a very large network or if you receive an error message warning you that the Registry is too small. You should be aware that setting a large value for the Registry size limit does not cause the system to use that much space unless it is actually needed by the Registry. Also be aware that a large value does not guarantee that the maximum space is actually available for use by the Registry.

To change Registry size limit value:

1. Launch Regedt32.

2. Select the Window menu option for **HKEY_LOCAL_MACHINE**.

3. Use the tree control in the left-hand window to navigate to the **SYSTEM\CurrentControlSet\Control** subkey. Click on the subkey to select it; its values appear in the right-hand window.

4. Click on the value entry **RegistrySizeLimit**.

5. On the Edit menu, select DWORD. The DWORD Editor appears.

6. **RegistrySizeLimit** must have a type of REG_DWORD and a data length of 4 bytes, or it will be ignored.

TIP: *If the RegistrySizeLimit value does not exist, it is because the default value has not been changed. You can create it using the technique in the "Adding New Registry Keys and Values" Immediate Solution earlier in this chapter or by setting a value via the graphical user interface. You control this setting in the GUI by using the System applet Control Panel. Once in the applet, select the Advanced tab, click on Performance Options, and then click on Change.*

Locating Windows 3.x Application INI Files Registry Keys

Not all of us can upgrade all of our applications every time a new version of Windows appears. Because of this, you might have to install 16-bit Windows 3.x applications on a Windows 2000 system. To troubleshoot these applications, you might have to locate and modify their INI file mapping information. Follow these steps to find where the Windows 2000 installation process has placed the INI file data for Windows 3.x applications:

1. Launch Regedt32.

2. Select the Window menu option for **HKEY_LOCAL_MACHINE**.

3. Use the tree control in the left-hand window to navigate to the **SOFTWARE\Microsoft\WindowsNT\CurrentVersion\ IniFile Mapping** subkey. Double-click on the subkey to open the software program subkeys under it.

4. You will see a set of subkeys for all the Windows 3.x programs that have been imported into the system. Click on the subkey in which you are interested and the path to its INI file is shown as a REG_SZ value. If you need to change the path, use the String Editor, as described in the "Modifying a Registry Value of String Data Type" Immediate Solution earlier in this chapter.

In string values that point to Registry nodes, there are several prefix characters that change the behavior of the INI file mapping, as shown in Table 1.2.

Table 1.2   INI file mapping symbols.

Symbol	Meaning
!	Forces all writes to go to the Registry and INI file on disk
#	Forces the Registry value to be set to the value in the INI file when a new user logs in for the first time following setup
@	Prevents reads from going to the INI file if the information is not stored in the Registry as well
USR	Stands for **HKEY_CURRENT_USER**
SYS	Stands for **HKEY_LOCAL_MACHINE\Software**

Locating MS-DOS Application Settings Information

MS-DOS applications also may be installed on a Windows 2000 machine. DOS applications do not use INI files, but there are a series of Registry values that control how they operate. To view the Registry values, follow these steps:

1. Launch Regedt32.

2. Select the Window menu option for **HKEY_LOCAL_MACHINE**.

3. Use the tree control in the left-hand window to navigate to the **SYSTEM\CurrentControlSet\Control\WOW** subkey. Click on the subkey to select it.

4. Click on the **wowcmdline** value entry.

5. On the Edit menu, select String. The String Editor appears.

6. You have several options for this entry, as summarized in Table 1.3. Information you enter is enclosed in **<>** (don't actually enter these symbols in the Registry). Make the appropriate changes and click on OK to save the information into the Registry.

TIP: *You can disable the 16-bit Windows On Windows subsystem or both the Windows On Windows subsystem and the Windows NT Virtual DOS Machine (NTVDM) subsystem. You can do this by placing a character in front of the strings for the cmdline and wowcmdline values. This change takes effect right away; it is not necessary to reboot. Adding a character to the beginning of the string allows you to easily reverse the change later to enable these subsystems. This change is often recommended when you are using Terminal Server because multiple instances of 16-bit applications can cause performance issues.*

Table 1.3   Entry values for wowcmdline and their effects.

Value	Effect
%systemroot%\System32\Ntvdm.exe	Specifies the default path and Virtual Device Manager (VDM) to run MS-DOS apps in
-a <vdmpath>	Specifies the command-line parameter for the current VDM
-f <path>	Changes the directory for Ntvdm.exe
-m	Hides the VDM console window display
-w <wowvdmpath>	Sets the Win16 on Win32 (WOW) VDM application

Locating Windows 2000 Security Settings

Windows 2000 has a substantial amount of security-related information in the Registry. Chapter 13 of this book details many security configurations we can make using the Windows Registry. Here, let's simply view these security settings. You should note how well protected this area of the Registry is thanks to very strong default NTFS (New Technology File System) security settings:

1. Launch Regedt32.

2. Select the Window menu option for **HKEY_LOCAL_MACHINE**.

3. Select the Security subkey in the left-hand window. Click on the Security menu item and choose Permissions. In the Permissions For Security window, select Allow Read for the Administrators group, then choose OK.

4. You can now double-click on the Security subkey to view the many security configurations stored in the Registry.

Locating Windows 2000 Software Package Settings

If you are a user or an administrator for a Windows 2000 network, sooner or later you will be confronted with the task of locating the Registry entries for an application package. Due to a specification from Microsoft for how this information is to be added to the Registry, you can follow these easy steps to find the entries for *any* application:

1. Launch Regedt32.

2. Select the Window menu option for **HKEY_LOCAL_MACHINE**.

3. Use the tree control in the left-hand window to navigate to the Software subkey. Double-click on the subkey to expand it.

4. Find the Registry subkey for the company that makes the application in question. For example, it would be "Microsoft" for Word version 7. Double-click on the subkey to open its subkeys.

5. Find the Registry subkey for the name of the application in which you are interested. Double-click on the subkey to open its subkeys.

6. Next, find the Registry subkey for the application version in which you are interested. Double-click on the subkey to open its subkeys and values.

7. You can now make the desired changes. They will take effect the next time you run the application (although you might have to reboot for some applications).

Locating Windows 2000 Registered File Extensions

Windows 2000 keeps Registry entries for file extensions that can trigger applications (for example, a .doc extension can automatically run Word to display the file). Here is how to access these settings directly within the Registry:

1. Launch Regedt32.

2. Select the Window menu option for **HKEY_LOCAL_MACHINE**.

3. Use the tree control in the left-hand window to navigate to the **SOFTWARE\Classes** subkey. Double-click on the subkey to select it and open its subkeys.

TIP: *You can access this same information via the **HKEY_CLASSES_ROOT** hive key.*

---

4. There are two kinds of subkeys under this entry: file extensions (noted with a period in front of them) and **COM GUID** values (covered in the next Immediate Solution). To find out which applications and menu commands are set up for a given extension, expand it and look through its value entries.

TIP: *This file association information can also be modified using the graphical user interface (GUI). Use the Tools To Folder Options menu selection in the Windows Explorer. Once there, access the File Types tab. This window provides a simple GUI interface for changing file associations.*

Locating Windows 2000 Registered COM Servers

Windows 2000 keeps Registry entries for all COM (OLE) servers. You can check this to find out where a given server's DLL or EXE application is located. Here is how to access these settings directly within the Registry:

1. Launch Regedt32.
2. Select the Window menu option for **HKEY_LOCAL_MACHINE**.
3. Use the tree control in the left-hand window to navigate to the **SOFTWARE\Classes** subkey. Double-click on the subkey to select it and open its subkeys.

*TIP: You can access this same information via the **HKEY_CLASSES_ROOT** hive key.*

4. To find out where a given COM server is located, expand its entry and look through its subkey entries for InProcServer32 or LocalServer32. A value entry under either of these keys gives the path and EXE or DLL name.

Locating Windows 2000 Hardware Settings

Perhaps the most important task the 2000 Registry performs is to connect the operating system with the hardware installed on a particular computer. To view the hardware entries in the Windows 2000 Registry (which are validated at each bootup of the machine), follow these steps:

1. Launch Regedt32.
2. Select the Window menu option for **HKEY_LOCAL_MACHINE**.
3. Use the tree control in the left-hand window to navigate to the HARDWARE subkey. Double-click on the HARDWARE subkey to select it and open its subkeys. There are three major subkeys for HARDWARE:

   • *Description*—Contains an entry for each hardware component. It has values for its version information, IRQ addresses, and a unique identifier. It can also have subkeys for other hardware elements or specific information regarding system hardware.

- *Devicemap*—Contains device driver file and system information, usually as REG_SZ values under subkeys for the specific devices.

- *Resourcemap*—Contains system resource information like IRQ (interrupt request) and port usage so that conflicts can be arbitrated. The information is stored in Binary format under subkeys for the various hardware elements.

Locating Windows 2000 Boot Information

Windows 2000 boots by using a complex process that is heavily dependent on the Registry. A specific set of Registry keys and subkeys is used in the arrangement that can be inspected and modified if problems occur. Here is how to view this set of keys and subkeys:

1. Launch Regedt32.

2. Select the Window menu option for **HKEY_LOCAL_MACHINE**.

3. Use the tree control in the left-hand window to navigate to the SYSTEM subkey. Double-click on the SYSTEM subkey to select it and open its subkeys. The following subkeys appear:

- *ControlSet001*—Contains the primary control set for Windows 2000. It is used by default to boot the system, but is copied to another CurrentControlSet first.

- *ControlSet002*—Contains the backup control set for Windows 2000. It is used to boot the system if ControlSet001 fails.

- *CurrentControlSet*—Contains the working control set for Windows 2000. It is used to actually boot the system.

- *CurrentControlSet\Control*—Contains a group of subkeys that define the state of the computer when it is booted.

- *CurrentControlSet\Control\ServiceGroupOrder*—Contains the list of groups (in a precise order) to load to construct the operating system. For example, the system must load the hard disk device drivers before the file system can be loaded.

- *CurrentControlSet\Services*—Contains all the device drivers used to boot the operating system.

Locating Windows 2000 Environment Variables

As mentioned in the In Brief section of this chapter, Environmental variables are used in Windows 2000 to allow scripts, Registry entries, and other applications to use wildcards for important system information that might change. This information is also presented in the System applet of Control Panel via the Advanced tab. To find the current list of Environmental variables and their values for Windows 2000 in the Registry, follow these steps:

1. Launch Regedt32.

2. Select the Window menu option for **HKEY_LOCAL_MACHINE**.

3. Use the tree control in the left-hand window to navigate to the **SYSTEM\CurrentControlSet\Control\Session Manager\ Environment** subkey. Double-click on the subkey to select it and open its information. Each Environmental variable has an entry. You can obtain Environmental variable names (which are used with other Registry entries as expandable strings) and values from these entries.

Locating Windows 3.x Default DLLs

Windows 2000 maintains a Registry entry for available DLLs it uses for Windows 3.x applications. To find out if a given DLL is on this list, follow these steps:

1. Launch Regedt32.

2. Select the Window menu option for **HKEY_LOCAL_MACHINE**.

3. Use the tree control in the left-hand window to navigate to the **SYSTEM\CurrentControlSet\Control\WOW** subkey. Double-click on the subkey to expand it.

4. Click on the **KnownDLLs** value entry.

5. On the Edit menu, select String. The String Editor appears. Each DLL known by the system and available for use by Windows 3.x applications is listed in this value entry, separated by spaces.

Locating Windows 2000 Current Control Information

When Windows 2000 boots, it ends up with a specific set of Registry entries that control how the operating system behaves. These are called *control sets*. To examine the control set that is currently in use for a given Windows 2000 computer, follow these steps:

1. Launch Regedt32.

2. Select the Window menu option for **HKEY_LOCAL_MACHINE**.

3. Use the tree control in the left-hand window to navigate to the **SYSTEM** subkey. Double-click on the subkey to select it and open its subkeys.

4. The **CurrentControlSet** subkey contains the working control set for Windows 2000. It is used to actually boot the system and remains in this entry throughout that session.

Enabling and Disabling Dr. Watson Debugging Utility

Windows 2000 ships with the well-known application *Dr. Watson*. Dr. Watson is a utility that catches all application program Win32 errors and logs them to a file. The behavior of this debugging utility is controlled within the Registry. To access the Dr. Watson Registry settings, follow these steps:

1. Launch Regedt32.

2. Select the Window menu option for **HKEY_LOCAL_MACHINE**.

3. Use the tree control in the left-hand window to navigate to the **SOFTWARE\Microsoft\WindowsNT\CurrentVersion\AeDebug** subkey. Double-click on the subkey to select it and display its values.

4. Select the **Auto** value and double-click on it to bring up its Editor dialog box. Enter a 0 to disable Dr. Watson or a 1 to enable it.

5. If your system has a different comparable utility, you can change the **Debugger** value to contain its name and command-line information also. For Dr. Watson, this value should be **drwtsn32 -p %ld -e %ld -g**.

TIP: *If you have Visual Studio installed, Dr. Watson will be replaced in the Debugger value with msdev.exe.*

Chapter 2

Registry Tools

In Brief

Although the Registry Editor is the tool to use to directly manipulate the Windows 2000 Registry, it is not the only tool available. Other tools include the Windows 2000 Microsoft Management Console (MMC), the Control Panel applets, user profiles, the Group Policies, and a command-line tool provided with the Windows 2000 CD-ROM. In this chapter you learn how to harness the power of the Registry with these tools.

Windows 2000 Microsoft Management Console (MMC)

The *Windows 2000 MMC* is an application that loads administrative consoles (provided by Microsoft and other vendors) that allow users to manage Windows 2000 systems. You can access an empty MMC from the Start menu by selecting Run and typing MMC in the Run dialog box. The Computer Management Console is a very common snap-in for the MMC. This utility is the primary administrative tool for Windows 2000 and is shown in Figure 2.1.

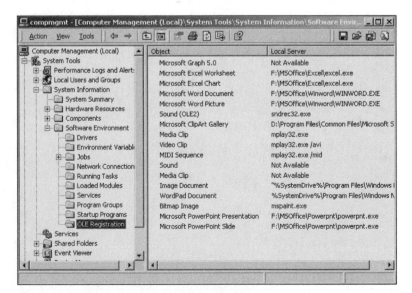

Figure 2.1   The Computer Management Console.

A simple way to access this console is by choosing Administrative Tools from the Start menu and selecting Computer Management. One of the entries in the Computer Management Console is System Information. As you might guess, the console builds its information from the **HKEY_LOCAL_MACHINE\HARDWARE and HKEY_LOCAL_MACHINE\SYSTEM\CurrentControlSet** keys and values. This includes information about the following:

- Services and device status
- Resources, such as IRQ (interrupt request), I/O ports, DMA (direct memory access), available memory, and specific device resources
- System and local user environments
- Network information, including general, transports, settings, and statistics
- Current memory status
- Drive status and statistics
- Display settings and status
- System status
- Windows 2000 version information

A nice feature of the System Information display in the Computer Management Console is that the information is read-only. It is intended to provide a quick inspection of Registry entries and to produce reports; you cannot use it to change Registry entries.

Control Panel Applets

There are 19 standard applets in the Windows 2000 Control Panel, all of which can affect Registry entries in some way. (Systems with additional hardware capabilities or installed software might have additional applets.) Figure 2.2 shows the main Control Panel display in Windows 2000 with the 19 standard applets. Table 2.1 provides a list of the standard applet names and capabilities. You may have more in your configuration due to additional installed components.

2. Registry Tools

Table 2.1   Control Panel applets and their effects in Windows 2000.

Applet Name	Affects these Type of Settings	Keys Manipulated
Accessibility Options	User	HKEY_CURRENT_USER\Control Panel
Add/Remove Hardware	System	HKEY_LOCAL_MACHINE\HARDWARE
Add/Remove Programs	System	HKEY_LOCAL_MACHINE\SOFTWARE
Administrative Tools	System	HKEY_LOCAL_MACHINE\SOFTWARE
Date/Time	System	HKEY_LOCAL_MACHINE\SOFTWARE
Display	System and User	HKEY_LOCAL_MACHINE\SYSTEM and HKEY_CURRENT_USER
Folder Options	System	HKEY_LOCAL_MACHINE\SOFTWARE
Fonts	System	HKEY_LOCAL_MACHINE\SOFTWARE
Internet Options	System	HKEY_LOCAL_MACHINE\SOFTWARE
Keyboard	System and User	HKEY_LOCAL_MACHINE\SOFTWARE and HKEY_CURRENT_USER
Mouse	System and User	HKEY_LOCAL_MACHINE\SOFTWARE and HKEY_CURRENT_USER
Network and Dial-Up	System	HKEY_LOCAL_MACHINE\SOFTWARE Connections
Phone and Modem	System	HKEY_LOCAL_MACHINE\SYSTEM Options
Power Options	System	HKEY_LOCAL_MACHINE\HARDWARE
Printers	System	HKEY_LOCAL_MACHINE\SYSTEM
Regional Options	System	HKEY_LOCAL_MACHINE\SYSTEM
Scheduled Tasks	System	HKEY_LOCAL_MACHINE\SOFTWARE
Sounds and Multimedia	System and User	HKEY_LOCAL_MACHINE\SOFTWARE and HKEY_CURRENT_USER
System	System and User	HKEY_LOCAL_MACHINE\SOFTWARE, HKEY_LOCAL_MACHINE\SYSTEM, and HKEY_CURRENT_USER

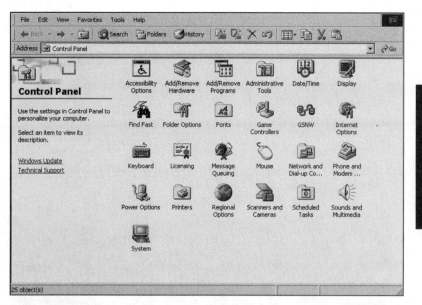

Figure 2.2   The Windows 2000 Control Panel.

The REG Utility

Microsoft includes some powerful and useful tools on the Windows 2000 CD-ROM called Windows 2000 Support Tools. These tools can be installed by double-clicking on the 2000RKST.MSI file located in the Support Tools subfolder. You should seriously consider installing all the Support Tools and their documentation if you support a Windows 2000 network. This section concentrates on one of these tools that is very handy for Registry work. This tool is called simply REG.

A Command-Line Utility with Powerful Options

REG is a command-line utility installed with the Windows 2000 Support Tools, as illustrated in Figure 2.3. It gives Windows 2000 administrators a number of powerful options, including the following:

- Displaying specific or general data from a Registry Key (**QUERY**)
- Adding new Registry Keys (**ADD**)
- Updating information in existing Registry Keys (**UPDATE**)
- Removing Registry Keys (**DELETE**)
- Copying Registry Keys from one location to another (**COPY**)
- Saving Registry Keys to hive files (**SAVE**)

- Restoring Registry Keys from hive files (**RESTORE**)
- Loading hive files into new Registry Keys (**LOAD**)
- Removing loaded hive files from Registry Keys (**UNLOAD**)
- Finding (and Replacing) Registry Keys or Values containing strings (**FIND**)
- Saving Registry Keys to REG files (**EXPORT**)
- Loading Registry Keys from REG files (**IMPORT**)
- Comparing Registry Keys on two machines or to a string and showing differences (**COMPARE**)

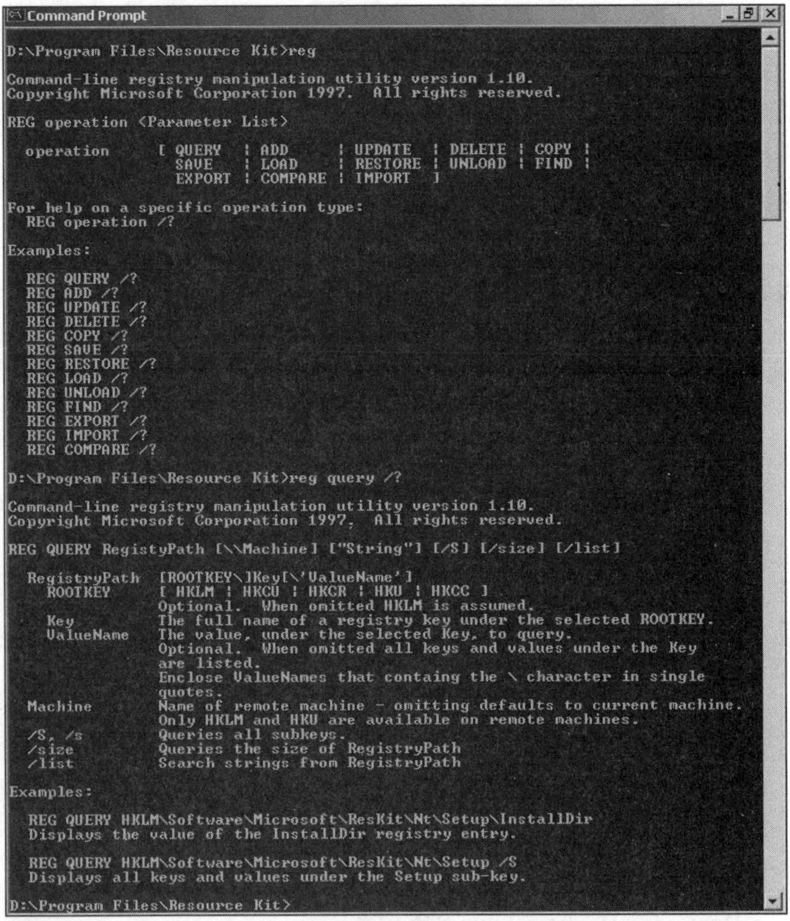

Figure 2.3   The Windows 2000 Support Tools **REG** application.

Fast Access to the Registry

REG is optimized for fast Registry access and thus can be much easier to work with than Regedt32. Also, because **REG** provides an automatic replacement feature for its **Find** command, it can do things that Regedt32 cannot. **REG** does not require the administrator to locate an entry before manipulating it, so it frees the administrator to do other work while the action is being performed.

Use in Batch Files

Because it is a command-line utility, **REG** can be used in batch files for automatic Registry updates, cleanups, and other administrative chores. Its outputs can also be redirected into text files or to the printer as needed, just like other command-line programs.

TIP: *The version of **REG** that you find in the Windows 2000 Support Tools has been dramatically enhanced compared to previous versions. If you have a library of existing batch files that use **REG**, you will have to check these files with the new version for appropriate syntax usage. Check the Windows 2000 Support Tools Help system for more information.*

REG Examples

Here are some examples of the power of the **REG** utility:

```
REG ADD HKLM\Software\MySoftwareCompany\
/v Path /t REG_EXPAND_SZ /d %"SystemRoot"%
```

This example adds a value to the Windows Registry in the **Software\ MySoftwareCompany** subkey. The value is **Path**, it is of type **REG_EXPAND_SZ**, and it contains the data **%"SystemRoot"%**.

```
REG COMPARE \\W2KSRV\HKLM\Software\MySoftwareCompany\
HKLM\Software\MySoftwareCompanyII /v Version
```

This example compares all of the **version** value under **HKLM\Software\ MySoftwareCompany** on the remote system named **W2KSRV** with the same value under **HKLM\Software\ MySoftwareCompanyII** on the current computer.

```
REG QUERY \\W2KSRV\HKLM\Software\Microsoft\ResKit /V version
```

This displays the contents of the Registry value **version** on the computer **\\W2KSRV**.

The System Policy Editor (SPE)

Windows 2000 provides the *System Policy Editor* (**Poledit.exe**) application to give administrators a simple user interface for making subtle but powerful changes to the system options available to various users. Some users can be allowed access to most configuration options, whereas others can be given virtually none. This permits many different users at many levels of security to use the same Windows 2000 computer. As you might guess, all of the changes made by SPE end up in the Registry.

Microsoft's new and improved System Policy Editor included with Windows 2000 is the Group Policy Editor. SPE might still be useful, however, in Windows environments not running Active Directory. See the next section of this chapter for more information. In order to check out the System Policy Editor, choose the Start menu, and then choose Run. In the Run dialog box, type **poledit** and choose OK.

The Group Policy Snap-In

If you are using Active Directory, the new directory service of Windows 2000, you can take advantage of a very powerful method for controlling users' environments via Registry changes. This new method is called *Group Policy*.

As an Active Directory Windows 2000 administrator, you assign Group Policies to users or computers in the network by using the Group Policy snap-in for the MMC. See the "Immediate Solutions" section in this chapter for the step-by-step instructions on how to do this.

Group Policy is the most powerful and simple method for controlling user environments that Microsoft has introduced so far. The main way to establish such control is to use the Administrative Templates sections of Group Policy. There is an Administrative Templates section for the User Configuration and the Computer Configuration in the Group Policy snap-in. These sections contain all the Registry-based Group Policy settings, including Windows Components, System, and Network. User Configuration settings are stored in the Registry in the **HKEY_CURRENT_USER** section, while the Computer Configuration settings are stored in **HKEY_LOCAL_MACHINE**. There are many settings stored in the two Administrative Template settings within Group Policy. In fact, there are more than 450 different settings for configuring the user environment alone.

Users in the Active Directory network can receive their settings from several Group Policy Objects (GPOs) that act together to configure their computer and desktop environments. This is because, by default, Group Policy settings flow down through the Active Directory hierarchy though a process known as *inheritance.*

Most of the time, you link GPOs to domains or organizational units, using the Active Directory Users and Computers snap-in. Group Policy Objects apply to a user or computer in the following order:

- *Local GPO*—Each Windows 2000 computer has a Local Group Policy Object assigned to it; this is the first to be processed.

- *Site GPOs*—Next, any Group Policy Objects that are linked to the site, the system, or the user, are located and processed. If there are multiple site GPOs, they process in the order that you specify.

- *Domain GPOs*—After that, any Group Policy Objects linked to the domain, the system, or the user, are located and are processed. If there are multiple Domain GPOs, they process in the order that you specify.

- *OU (Organizational Unit) GPOs*—Finally, any GPOs linked to OUs above the GPO that contain the user or computer are processed. These are called parent GPOs. Also, any GPOs linked to the OU, the system, or the user, are located and are processed. If there are multiple OU GPOs, they process in the order that you specify.

RegMon

Monitoring changes to the Registry is an excellent idea—both from a security perspective and in determining what change is made when you use the graphical user interface for management. You would really have your work cut out for you if you were trying to use Regedt32 to do this. There is an excellent tool called **RegMon** that allows you to easily monitor such changes to the Registry. This program is available from **www.sysinternals.com**.

RegClean

Many applications write changes to the Registry and leave things behind when you remove those applications. Trying to clean your Registry of unwanted or unnecessary entries yourself can be a pretty daunting task. Fortunately, Microsoft provides a tool for doing automated Registry housecleaning: **RegClean**.

RegClean is available from **ftp://ftp.microsoft.com/Softlib/ MSLFILES**. Be sure to read the *ReadMe.txt* file for any special instructions for installing and using the product successfully. See the "Immediate Solutions" section of this chapter for step-by-step guidance on using **RegClean**.

RegMaid

For a much more interactive version of **RegClean**, you might want to investigate **RegMaid**, which is also available at ftp:**//ftp.microsoft.com/ Softlib/MSLFILES**. For step-by-step instructions on its installation see the "Immediate Solutions" section of this chapter. This tool analyzes the Registry as **RegClean** does, while providing the advanced user with an interface for viewing potential problems.

The CLSID view is shown when **RegMaid** is first started because this is the view where the cleanup process typically starts. To populate this view **RegMaid** scans the Registry subkeys in **HKEY_ CLASSES_ROOT\CLSID**. In each CLSID entry, **RegMaid** looks for at least one of the following entry keys: InProcHandler, InProcHandler32, InProcServer, InProcServer32, LocalServer, or LocalServer32. For each of these entries found, **RegMaid** expects an associated file name. If that file name has a fully qualified path, then that is where **RegMaid** looks for the file; otherwise **RegMaid** searches the system paths just as if it were a client trying to find the server. If any entry in a CLSID key has a file that **RegMaid** could not find, then that CLSID key is considered problematic and listed in this view.

You can use the The ProgId View button on the toolbar to switch to the ProgId view. This view is associated with the ProgId entry keys in **HKEY_CLASSES_ROOT**. **RegMaid** takes the GUID associated with this CLSID subkey and tries to find a matching subkey in **HKEY_ CLASSES_ROOT\CLSID**. If a match cannot be found, the ProgId entry is considered problematic and added to the ProgId view.

The TypeLib View button switches you to TypeLib view. To populate this view **RegMaid** scans the Registry subkeys in **HKEY_ CLASSES_ROOT\TypeLib** to see if they contain a subkey associated with a .TLB file. If they do, **RegMaid** uses the same technique as in the CLSID view to find the file. If the .TLB file cannot be found, the TypeLib entry is added to the list.

The Interface View button is used to access the Interface view. **RegMaid** makes its evaluation of the Interface entries in the same way it does for the ProgId entries. It searches the **HKEY_CLASSES_ ROOT\Interface** for keys that contain a TypeLib subkey. When a TypeLib subkey is found, **RegMaid** tries to find a match in **HKEY_CLASSES_ROOT\TypeLib**. If **RegMaid** cannot find a match, it adds the Interface key to this view.

Immediate Solutions

Creating a System Report from Registry Entries with Computer Management

The Computer Management Console is an MMC snap-in that compiles a sizable group of Registry entries and displays them using an Explorer-like user interface. To use this snap-in to generate reports about the current Registry for a Windows 2000 computer, follow these steps:

1. Start the Computer Management Console from the Start menu (via Programs|Administrative Tools|Computer Management). The MMC Explorer-style user interface appears.

2. Select the System Information element in the left-hand tree control.

3. Click on the Action menu item; a dropdown menu appears.

4. As shown in Figure 2.4, you have two choices to make when generating your Registry report: You can have it output in the NFO system information format (which can be read only by another Windows 2000 MMC application) or in a text file (which can be viewed in Notepad or any other word processor).

5. No matter which option you choose, a file browser dialog box appears for you to either select an existing file or create a new one. Click on OK when you are finished. You are prompted for any needed additional information, and the report is generated and output in your chosen format.

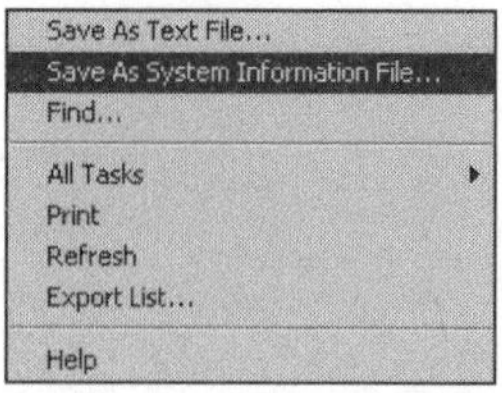

```
Save As Text File...
Save As System Information File...
Find...

All Tasks                      ▶
Print
Refresh
Export List...

Help
```

Figure 2.4   The MMC System Information Action menu options.

Changing the Recovery Utility's Registry Entries with Control Panel

Windows 2000 offers a Recovery utility that takes over in the event of a system lockup or crash. Follow these steps to use Control Panel to configure the Recovery utility's behavior:

1. On the Start menu, select Settings|Control Panel. The Control Panel window appears.

2. Double-click on the System applet. Select the Advanced tab (see Figure 2.5) and then click on the Startup And Recovery options button.

3. Figure 2.6 shows the property sheet that appears. (If you have a dual or multiboot configuration, the dropdown list of operating systems is displayed as shown in Figure 2.6; otherwise, it won't.) Make any desired changes to the Recovery utility settings and click on OK to save the values to the Registry. The available checkboxes are:

   - *Write An Event To The System Log*—Controls a REG_DWORD entry. If this checkbox is checked, the entry has a value of 1; otherwise, it has a value of 0. This entry controls whether events are written to a system log for later examination. The default values are 1 for Server and 0 for Workstation.

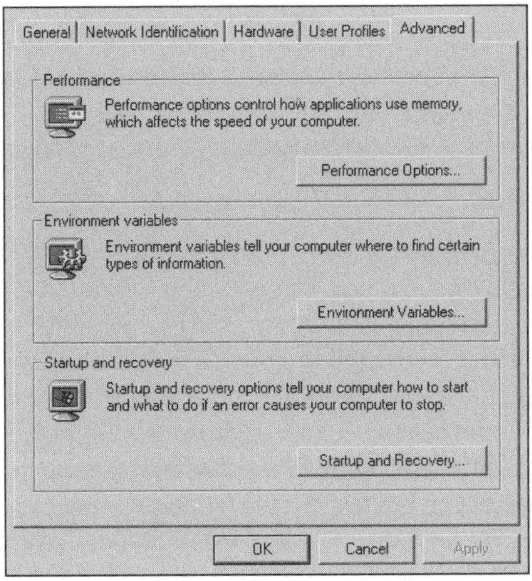

Figure 2.5   The Advanced tab in the System applet in Control Panel.

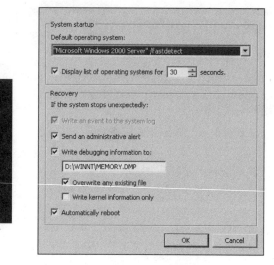

Figure 2.6   The Recovery options on the Startup/Shutdown dialog box.

- *Send An Administrative Alert*—Controls a REG_DWORD entry. If this checkbox is checked, the entry has a value of 1; otherwise, it has a value of 0. This entry controls whether an error message is displayed when Overwrite is 0 and LogEvent is 1 and the log file runs out of space. All logging terminates after the error message is sent, but the system keeps running. The default values are 1 for Server and 0 for Workstation.

TIP: *You can alter the Send An Administrative Alert default behavior by creating a new Registry value at **HKEY_LOCAL_MACHINE\SYSTEM\CurrentControlSet\Control\Lsa** called **CrashOnAuditFail**. It should be of type REG_DWORD and set to 1. When a situation occurs that normally causes a reboot, this entry causes the system to crash and not reboot.*

- *Write Debugging Information To*—Controls a REG_DWORD entry. If this checkbox is checked, the entry has a value of 1; otherwise, it has a value of 0. This entry controls whether a log file is created when a lockup happens. The default values are 1 for Server and 0 for Workstation. The edit control under the Write Debugging Information To checkbox enters a REG_SZ entry that is the file name and pathname for the crash dump information from the Write Debugging Information To checkbox. The default is %systemroot%\Memory.dmp.

- *Overwrite Any Existing File*—Controls a REG_DWORD entry. If this checkbox is checked, the entry has a value of 1; otherwise, it has a value of 0. This entry controls whether a log file is overwritten when new information is entered or the new information is appended to an existing file. The default values are 1 for Server and 0 for Workstation.

- *Write Kernel Information Only*—Causes only the information necessary for kernel debugging to be written to the file.

- *Automatically Reboot*—Controls a REG_DWORD entry. If this checkbox is checked, the entry has a value of 1; otherwise, it has a value of 0. This entry controls whether an automatic reboot happens when a failure occurs. The default values are 1 for Server and 0 for Workstation.

Setting the Display's Color Registry Entries with Control Panel

The *Display applet* in Control Panel lets you configure the color settings for 29 of the most important aspects of the Windows 2000 user-interface elements. These settings are then kept in the Registry for the currently logged-on user and reset each time the user logs on.

Table 2.2 shows this applet's 29 available user-interface elements and their default settings in the Registry. You can change these values by using the graphical controls of the applet or entering the desired numbers directly. Each is a REG_SZ value and consists of three numbers representing a red, green, and blue color component ranging from 0 (black) through 255 (pure red, green, or blue).

Table 2.2   User-interface color settings in the Display applet in Control Panel.

User-Interface Element	Default Values	Color
ActiveBorder	212 208 200	Blue-gray
ActiveTitle	10 36 106	Dark blue
AppWorkSpace	128 128 128	Light gray
Background	58 110 165	Off-white
ButtonAlternateFace	181 181 181	Medium gray
ButtonDkShadow	64 64 64	Dark gray

(continued)

Table 2.2   User-interface color settings in the Display applet in Control Panel *(continued)*.

User-Interface Element	Default Values	Color
ButtonFace	212 208 200	Light gray
ButtonHilight	255 255 255	White
ButtonLight	212 208 200	Light gray
ButtonShadow	128 128 128	Medium gray
ButtonText	0 0 0	Black
GradientActiveTitle	166 202 240	Light blue
GradientInactiveTitle	192 192 192	Light gray
GrayText	128 128 128	Medium gray
Hilight	10 36 106	Dark blue
HilightText	255 255 255	White
HotTrackingColor	0 0 128	Bright blue
InactiveBorder	212 208 200	Light gray
InactiveTitle	128 128 128	Light gray
InactiveTitleText	212 208 200	Gray
InfoText	0 0 0	Black
InfoWindow	255 255 255	White
Menu	212 208 200	Gray
MenuText	0 0 0	Black
Scrollbar	212 208 200	Light gray
TitleText	255 255 255	White
Window	255 255 255	White
WindowFrame	0 0 0	Black
WindowText	0 0 0	Black

Here's how to configure the color settings for Windows 2000 user-interface elements:

*TIP: The color settings are found in the Registry at **HKEY_CURRENT_USER\Control Panel\Colors**. Each user-interface element is a value with a REG_SZ data type.*

1. On the Start menu, select Settings|Control Panel. The Control Panel window appears.

2. Double-click on the Display applet. Select the Appearance tab, as shown in Figure 2.7. Make any desired changes and click on OK to save the values to the Registry.

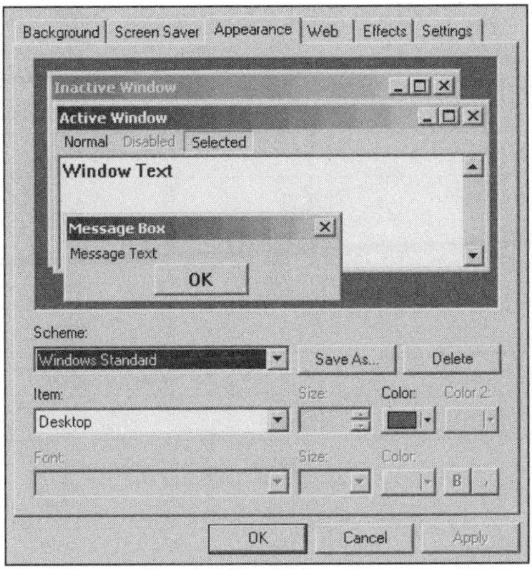

Figure 2.7   The Appearance tab in the Display applet in Control Panel.

TIP: Pure black is 0 0 0. Any color with all three numbers the same is a shade of gray; the higher the numbers, the lighter the shade. 255 255 255 is pure white.

Changing the Desktop Color Settings in the Registry with Control Panel Display Schemes

One of the options in the Display applet is to use custom colors for the various user-interface elements. This requires using the Display applet's user interface rather extensively, which can be time consuming. However, you can make this process much easier by using the Display applet in the Control Panel feature of desktop color schemes. Table 2.3 shows the 36 default color schemes available in the Display applet and their overall appearance, font size, and any special attributes.

Table 2.3   Display scheme names and settings in the Display applet.

Name	Theme	Font Type	Special
Brick	Reddish	Standard	N/A
Desert	Yellowish	Standard	N/A
Eggplant	Brown/purple	Standard	N/A

(continued)

2. Registry Tools

Table 2.3   Display scheme names and settings in the Display applet *(continued)*.

Name	Theme	Font Type	Special
High Contrast #1	Yellow/black	Standard	Mono laptops
High Contrast #1 Extra Large	Yellow/black	V Large	Mono laptops
High Contrast #1 Large	Yellow/black	Large	Mono laptops
High Contrast #2	Green/black	Standard	Mono laptops
High Contrast #2 Extra Large	Green/black	V Large	Mono laptops
High Contrast #2 Large	Green/black	Large	Mono laptops
High Contrast Black	White/black	Standard	Mono laptops
High Contrast Black Extra Large	White/black	V Large	Mono laptops
High Contrast Black Large	White/black	Large	Mono laptops
High Contrast White	Black/white	Standard	Mono laptops
High Contrast White Extra Large	Black/white	V Large	Mono laptops
High Contrast White Large	Black/white	Large	Mono laptops
Lilac	Purplish	Standard	N/A
Lilac Large	Purplish	Large	N/A
Maple	Brownish	Standard	N/A
Marine High Color	Bluish	Standard	Needs 24-bit VGA
Plum High Color	Blue/gray	Standard	Needs 24-bit VGA
Pumpkin Large	Yellowish	Large	N/A
Rainy Day	Dark Blue	Standard	N/A
Red White And Blue VGA	Grayish	Standard	16-color VGA
Rose	Pinkish	Standard	N/A
Rose Large	Pinkish	Large	N/A
Slate	Blue/gray	Standard	N/A
Spruce	Greenish	Standard	N/A
Storm VGA	Gray/blue	Standard	16-color VGA
Teal VGA	Gray/aqua	Standard	16-color VGA
Wheat	Yellow/green	Standard	N/A
Windows Classic	Gray/blue	Standard	N/A
Windows Classic Extra Large	Gray/blue	Standard	N/A
Windows Classic Large	Gray/blue	Standard	N/A
Windows Standard	Gray/blue	Standard	N/A
Windows Standard Extra Large	Gray/blue	Standard	N/A
Windows Standard Large	Gray/blue	Standard	N/A

To change the desktop color scheme in the Registry using the Display applet, follow these steps:

TIP: *The color scheme settings are found in the Registry at **HKEY_CURRENT_USER\Control Panel\Appearance\Schemes**. You can create your own color schemes by adding values to this key, but this is difficult because the values are in Binary.*

---

1. On the Start menu, select Settings|Control Panel. The Control Panel window appears.

2. Double-click on the Display applet. Select the Appearance tab. Select the desired color scheme, and click on OK to save the values to the Registry. This sets all the various color elements of the display at one time, rather than requiring tedious manual configurations. Figure 2.8 shows the Maple display scheme selected in the Appearance tab.

Related solution:	Found on page:
Setting Screen Background Color	100

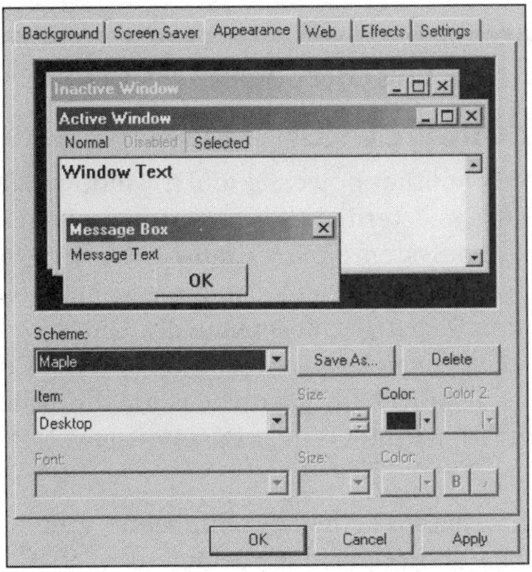

Figure 2.8   The Appearance tab of the Display applet in Control Panel showing a preset color scheme.

Using **REG** to Query the Registry

In order to take advantage of this "Immediate Solution," be sure you have installed the **REG** utility. **REG** is part of the Windows 2000 Support Tools included on the CD-ROM. See the **REG** Utility section of In Brief in this chapter for installation instructions.

1. Click on Start to Run.

2. In the Run dialog box, type **CMD** and click on OK.

3. In the Command Prompt window, type the following to query the contents of the Software subkey of **HKEY_LOCAL_MACHINE** and write these contents to a text file named **test.txt**:

```
REG QUERY hklm\software >c:\test.txt
```

4. Type **EXIT** and press Enter to close the Command Prompt window.

Setting Control Panel Display Icon Access with the System Policy Editor

The System Policy Editor is another powerful tool a Windows 2000 system administrator can use to determine how much control a given user has over a desktop. The System Policy Editor is a utility that permits setting many Registry permissions using a graphical user interface rather than setting Registry values manually. One very effective use for a system policy is in limiting how much of the Control Panel's Display Properties sheet any user who is not the system administrator can access. To set the Control Panel display, follow these steps:

1. Start the System Policy Editor via the Start|Run menu, then, enter **poledit**. Figure 2.9 shows the SPE's default user interface. Open the current POL file for the system via the File|Open Policy menu option.

2. Double-click on the appropriate User icon. In the Property dialog box, scroll down to the Control Panel entry. Expand the entry until you see the Display|Restrict Display entry. This entry has checkboxes for the items described next. The

Figure 2.9   The System Policy Editor in Windows 2000.

checkboxes are grayed out until you check the Restrict
Display checkbox. Checking a checkbox in the bottom portion
of the dialog box enables the Restrict Display restriction;
unchecking a checkbox disables the restriction. The available
restrictions are as follows:

- *Hide Settings tab*—Settings tab not shown if checked.

- *Hide Appearance tab*—Appearance dialog box not available
  if checked.

- *Hide Background tab*—Background dialog box not available
  if checked.

- *Hide Screen Saver tab*—Screen Saver tab not shown if
  checked.

- *Deny Access To Display icon*—Entire Display Properties
  sheet not available if checked.

3. Choose the desired restrictions, then click on OK to set the
   restrictions.

Enabling Shell Restrictions with the System Policy Editor

Windows 2000 system administrators can use the System Policy Editor to limit how much of the Windows 2000 shell any user (other than the system administrator) can access. To limit Windows 2000 shell access, follow these steps:

1. Start the System Policy Editor via the Start|Run menu; then, enter **poledit**. Open the current POL file for the system via the File|Open Policy menu option.

2. Double-click on the appropriate User icon. In the property dialog box, scroll down to the Shell entry. Expand the entry until you see the Restrictions entry with checkboxes for the items described next. Checking an item's checkbox enables the restriction; unchecking it disables the restriction. The following list describes how each option restricts user access when the item is selected:

   - *Remove Run command*—Disables the *Run* option on the Start menu.

   - *Remove Folders*—Removes Folder Changes entry from Settings menu.

   - *Remove Taskbar*—Hides the taskbar.

   - *Remove Find command*—Disables the **Find** command on the Start menu.

   - *Hide Drives*—Indicates which drives are not shown (1=A, 3=C, and so forth).

   - *Hide Network Neighborhood*—Hides the Network Neighborhood icon.

   - *No Entire Network In Network Neighborhood*—Shows only the network accessible by the user.

   - *No Workgroup In Network Neighborhood*—Hides local areas that are not accessible by the user.

   - *Hide All Items On Desktop*—Disables shortcut icons on the desktop.

   - *Remove Shutdown Command From Start Menu*—Disables the Shutdown option on the Start menu.

   - *Don't Save Settings At Exit*—Discards user profile changes when a user logs off.

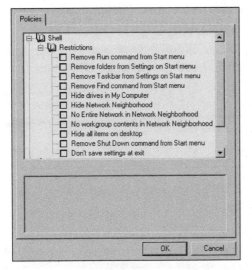

Figure 2.10   The System Policy Editor Shell Restrictions property sheet in Windows 2000.

3. Specify the restrictions you want to implement, and then click on OK to set the restrictions. Figure 2.10 gives an illustration of the property sheet with no restrictions set.

Using the Group Policy Editor

If you are using Active Directory with Windows 2000, your graphical user interface method for controlling user's environments is the Group Policy Editor, as follows:

1. In the Start menu, choose Run.

2. In the Run dialog, type MMC and click on OK.

3. Select the Console menu and choose Add/Remove Snap-in.

4. Click on the Add button on the Standalone tab and choose Group Policy from the list of snap-ins.

5. To edit the Group Policy Object associated with the local computer choose Local Computer and click on Finish.

6. Click on OK in the Add/Remove Snap-in window.

Explore the categories of Group Policy Object settings that you can make. Allow some time for this, as there are literally hundreds of settings that you can control.

Using **RegMon** to Monitor the Registry

In order to use this "Immediate Solution," be sure you install the **RegMon** utility as described in the **RegMon** "In Brief" section of this chapter.

1. Launch the **RegMon** utility. **RegMon** starts in capture mode.

2. In the Edit menu, choose the Filter/Highlight command.

3. Use this dialog to filter the Registry transactions you would like to capture.

4. Use the **File - Save** and **File - Save As** commands to save the captured Registry transactions.

Using **RegClean**

RegClean does a great job of saving undo information to the directory from which it is executed, so remember this in the event that **RegClean** inadvertently causes problems for you with its Registry manipulations. Here is how easy it is to use **RegClean**:

1. Use the Windows Explorer to double-click on the **RegClean.exe** file that you downloaded. This runs a self-extraction utility. Specify a directory to which you want to unzip **RegClean** (for example, c:\regclean) and click on Unzip.

2. Use the Windows Explorer to navigate to the directory you created and double-click on the **RegClean.exe** file.

3. **RegClean** launches and begins checking the Registry for errors (entries that have been left behind or added unnecessarily).

4. Once **RegClean** has finished checking the Registry, use the Fix Errors button to have **RegClean** work its magic.

5. Remember, **RegClean** creates a file with a name that starts with **UNDO** so you can easily undo its recommended changes.

Using **RegMaid**

RegMaid offers many advanced features that you can call upon as a Windows 2000 administrator. Here is how easy it is to install and use:

1. Use the Windows Explorer to double-click on the regmaid.exe file you downloaded. This runs a self-extraction utility. Specify a directory to which you want to unzip **RegClean** (for example, c:\regmaid) and click on Unzip.

2. Use the Windows Explorer to navigate to the directory you created, and double-click on the **RegMaid.exe** file located in the Release directory.

2. Registry Tools

Registry Disaster Prevention

In Brief

You should realize just how important the Registry is for a Windows 2000 system. It really is the central nervous system for the whole Windows 2000 machine. Manually manipulating the Registry can be very rewarding for you as a Windows 2000 administrator, but it can also be very dangerous. Fortunately, there are many methods for protecting your Registry information and for making sure you can recover from any crisis you might inadvertently cause with your manipulations. This chapter presents you with many solutions for safeguarding the most important part of your Windows 2000 computer.

Startup Options

Windows 2000 features a robust set of startup options that you can use to assist in troubleshooting if your system is having problems. You should be aware of all of these options and how to use each in troubleshooting. You access these startup options by pressing F8 when prompted at system startup. This occurs before the graphical portion of the Windows 2000 startup process.

- *Safe Mode*—This option starts Windows 2000 using basic files and drivers and does not load devices necessary to support networking.

- *Safe Mode With Networking*—This option starts Windows 2000 using basic files and drivers only and also includes networking support.

- *Safe Mode With Command Prompt*—This option starts Windows 2000 using basic files and drivers only, without support for networking. Once Windows 2000 is loaded, this option displays only the command prompt.

- *Enable Boot Logging*—This option creates a boot log of devices and services that are loading. The log is saved to a file named *Ntbtlog.txt* in the system root. All of the options listed above create this log by default.

- *Enable VGA (Video Graphics Array) Mode*—This option starts Windows 2000 using the basic VGA (video) driver. This mode is useful when you have installed a new driver for your video card that prevents Windows 2000 from starting properly. The basic video driver is also used by default when you start Windows 2000 in any kind of Safe Mode.

- *Last Known Good Configuration*—This option starts Windows 2000 using the Registry information that Windows saved upon the last successful shutdown.

- *Directory Services Restore Mode*—This option restores Active Directory on a domain controller.

- *Debugging Mode*—This option starts Windows 2000 and sends debug information through a serial cable to another computer.

The Windows 2000 Recovery Console

New to Windows 2000 is a disaster prevention device called the Recovery Console. The *Recovery Console* permits access to the file system without actually running Windows 2000 (W2K). This is obviously critical in the event that W2K is so badly damaged that the system cannot start. Using this important console, you can start and stop services, manipulate files and folders, and even repair broken boot sectors. You can even protect the Recovery Console from unauthorized usage by requiring administrator credentials. See the "Immediate Solutions" section of this chapter for information on installing and running the Recovery Console. Table 3.1 provides the commands supported by this incredible new tool.

Table 3.1   Recovery Console commands.

Command	Function
Attrib	Changes the attributes of a folder or file
Batch	Executes a batch file
CD	Changes the current folder or displays the current volume
Chkdsk	Checks the hard disk and repairs volumes if necessary
Cls	Clears the screen
Copy	Copies a single file to a specific location
Del	Deletes a file
Dir	Displays a list of a folder's contents
Disable	Disables a service or driver
Diskpart	Manages the partitions on your hard disk
Enable	Enables a service or driver

(continued)

Table 3.1   Recovery Console commands *(continued)*.

Command	Function
Exit	Quits the Recovery Console and restarts the system
Expand	Uncompresses a file from the CD-ROM or from a CAB (cabinet) file and copies it to the hard drive
Fixboot	Rewrites the boot sector code on the disk
Fixmbr	Rewrites the master boot record of the boot hard disk
Format	Formats a specified volume
Help	Shows help for Recovery Console commands
Listsvc	Lists available services and drivers
Logon	Logs on the administrator; prompts for the correct password
Map	Displays boot.ini drive letter mappings
MD	Creates a directory
More	Displays a text file
Rd	Deletes a directory
Ren	Renames a file or directory
Set	Displays and sets Recovery Console environment variables
Systemroot	Sets the current directory to the %systemroot% directory

The Emergency Repair Disk (ERD)

Many of us are accustomed to using the Emergency Repair Disk in Windows NT to repair the Registry. While the ERD of Windows 2000 does not contain a copy of the Registry to be used in repairs, it can perform important related functions. Therefore, you should be sure to make an Emergency Repair Disk and keep it up to date.

The *Windows 2000 Emergency Repair Disk* ensures you have a backup copy of the configuration files used to run DOS and 16-bit Windows programs and also the current setup log file. The disk also contains a copy of the default user profile.

Windows NT Backup

Windows 2000 offers a new and improved NT backup utility called **NTBACKUP.EXE** (see Figure 3.1). This utility allows you to back up the Registry and other key configuration components by simply backing up a single setting called system state data. System state data actually consists of several key elements, including the following:

- The Registry

- COM+ (Component Object Model+) Class Registration database

- The boot files
- Certificate services database
- Active Directory
- SYSVOL
- Cluster service information

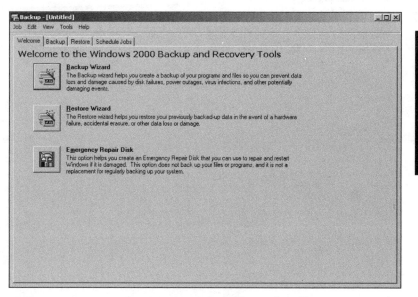

Figure 3.1   The new backup application in Windows 2000.

REGBACK and REGREST

The old Windows NT 3.1 Resource Kit introduced these classic utilities for NT, and they still work just fine in the Windows 2000 world. In fact, they have been showing up in many Resource Kits and supplements ever since. As its name implies, **REGBACK** is a command-prompt utility for performing Registry backups and **REGREST** is the companion restore utility.

While **REGBACK** might seem much too primitive compared to the new **NTBACKUP.EXE**, there might be reasons for keeping it around. It is, for example, very useful for backing up to a variety of different media. It is also very handy for those with a passion for command prompt sessions and batch files. In addition, **REGBACK** stores its backups as simple uncompressed files that do not need any special software for analysis and utilization. In fact, they can be used to re-create a registry on another system in the event of a real emergency.

Regedt32

The Registry Editor can also be used to do simple backups and re-stores of the Registry, compliments of a very handy Save feature that permits you to save a portion of the Registry as a text file. As you might guess, Regedt32 also lets you import these saved files.

Immediate Solutions

Creating a Set of Windows 2000 Setup Disks

You will notice that many of the disaster prevention Immediate Solutions in this chapter require you to start the Windows 2000 system from the CD-ROM or the Windows 2000 setup disks. This solution provides you with a convenient set of instructions for creating these setup disks:

1. Label four blank, formatted floppy disks as follows:

    - Windows 2000 Setup Boot Disk

    - Windows 2000 Setup Disk #2

    - Windows 2000 Setup Disk #3

    - Windows 2000 Setup Disk #4

2. Insert the first blank, formatted 1.44MB disk into the floppy disk drive of your Windows 2000 system.

3. Insert the Windows 2000 CD-ROM.

4. Click on Start, then click on Run.

5. In the Run dialog type:

```
d:\bootdisk\makebt32 a:
```

Where **d:** is the drive letter assigned to your CD-ROM drive.

6. Click on OK.

Creating a Windows 2000 Startup Floppy Disk

You should always keep a Windows 2000 startup floppy disk in your disaster prevention toolkit. This disk allows you to boot the system and access a drive with a faulty boot sequence. To create the disk, follow these steps:

1. Insert a blank, formatted disk into your Windows 2000 system.

2. Insert the Windows 2000 CD-ROM into the system.

3. Copy NTLDR (NT Loader), **Ntdetect.com**, and boot.ini from the Windows 2000 CD-ROM to the floppy disk.

4. If you have a SCSI system and the SCSI BIOS is not enabled, copy Ntbootdd.sys to the floppy disk.

TIP: *Be sure to recreate your Windows 2000 startup floppy disk after the installation of any Windows 2000 Service Pack. If you are making a Windows 2000 startup floppy disk while your computer is still functional, copy boot.ini from the hard disk drive. If you have not yet made a Windows 2000 startup floppy disk and you are encountering problems, you should copy the necessary files from another Windows 2000 computer and modify the boot.ini to match the configuration of your Windows 2000 system.*

Starting the Recovery Console

Follow these steps to start the Recovery Console on your system:

1. Start the system using the Windows 2000 CD-ROM or the Windows 2000 setup disks.

2. Enter Windows 2000 Setup.

3. Press Enter at the Setup Notification screen.

4. Press R to repair a Windows 2000 installation, and then press C to use the Recovery Console.

5. The Recovery Console displays valid Windows 2000 installations and prompts you to select the installation to repair. To access the disk with the Recovery Console, press the number key for the Windows 2000 installation that you want to repair, and then press Enter.

Installing the Recovery Console on a System

One powerful option is to install the Recovery Console on your Windows 2000 system so it can be easily accessed from the boot menu without requiring the Windows 2000 CD-ROM or setup disks. You should note that if your master boot record or system boot record needs repair, you will still have to access the Recovery Console from the CD-ROM or setup disks. You should also note that this cannot be accomplished if you are using a mirrored volume. To install the Recovery Console on your system, follow these steps:

1. Insert the Windows 2000 CD-ROM.

2. Select Start, then choose Run.

3. In the Run dialog, type:

```
d:\I386\Winnt32.exe /cmdcons
```

where **d:** represents the CD-ROM drive.

Using the Recovery Console to Restore the Registry

You can use the Recovery Console to easily restore a copy of the Registry. It is especially easy if you frequently back up the Registry using the Immediate Solution in this chapter called "Creating an Emergency Repair Disk." Follow these instructions if that is the case:

1. Start the Recovery Console, and log on to the Windows 2000 installation with the Registry that you want to restore.

2. Copy the files you want from %systemroot%\Repair\RegBack to %systemroot%\System32\Config, using the command

```
cd repair\regback
```

and

```
copy file_name drive_letter:\system_root\system32\config
```

where **file_name** is the Registry file you want to restore, **drive_letter** is the drive letter where your system is installed, and **system_root** is the system installation folder. To restore your entire Registry, copy the files Default, Sam, Security, Software, and System.

3. Exit the Recovery Console by typing **EXIT**.

3. Registry Disaster Prevention

Creating an Emergency Repair Disk

While the Emergency Repair Disk no longer contains a copy of the Registry, it can be quite useful in recovering a damaged system. It does have the ability to back up the Registry to a special directory on the system. Here are the steps to creating the ERD and creating the backup Registry copy:

1. Select Start to Run, type **NTBACKUP**, and click on OK.

2. Click on the Emergency Repair Disk button.

3. In the Emergency Repair Diskette window (see Figure 3.2), be sure to select the checkbox labeled, "Also back up the Registry to the repair directory. This backup can be used to help recover your system if the Registry is damaged."

This Registry Backup option causes Windows 2000 Backup to copy the registry hives to %systemroot%\repair. You can then use Windows 2000 Backup to back up the files from that location or manually move them to a safe location yourself.

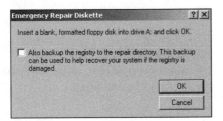

Figure 3.2   The Emergency Repair Diskette window.

Restoring Your System with the Emergency Repair Disk

To restore your system with the Emergency Repair Disk, perform the following:

1. Use the Windows 2000 CD-ROM or the setup disks to start your system.

2. At the Setup Notification screen, press Enter to continue.

3. At the Welcome To Setup screen, press R to select the option to repair a Windows 2000 installation.

4. Press R to repair a Windows 2000 installation by using the Emergency Repair Process.

5. Choose the type of repair option you want to use:

   • Press M for Manual Repair.

   • Press F for Fast Repair. Fast Repair requires no user interaction, while Manual Repair does.

6. Follow the instructions and insert the Emergency Repair Disk when prompted.

Backing up the Registry "Manually"

Remember that the Registry configuration is stored on your hard disk in a series of files located in %systemroot%\system32\config, which makes it possible to do "manual" backups. The challenge is, however, that when you are currently in Windows 2000, the operating system has these files in use so your manipulations of them are very limited. This can be overcome, though, and there are several possibilities and things to think about:

• You can use the Recovery Console to boot the system and copy the files to an alternate location and then copy them to a tape drive or removable media. This extra step is necessary because the Recovery Console does not permit the loading of additional drivers for removable media and assorted gadgets.

• If you have another operating system loaded on the machine (you are in a dual-boot configuration), you can boot to the alternate operating system and copy the files to the alternate media.

• You could use a DOS, Linux, or OS/2 boot disk that includes a command prompt shell and boot the system with this disk. If you are using the NTFS (New Technology File System) file system, you will need special drivers in order to read from this file system. You obviously use the command prompt shell to perform the file manipulations.

• If you are trying to copy the files to a floppy disk, you need some type of compression software, because the files that make up the Registry are going to be several megabytes in size at least.

Once you have successfully booted the system without Windows 2000 having a grasp on the Registry files, you are ready to "manually" back up the files. Follow these steps:

1. Boot your computer using one of the methods above and navigate to the %systemroot%\system32\config directory. This directory contains the files that make up your Registry and associated logs. Be sure you can actually see these files. On our sample Windows 2000 Server, this amounted to 28 files at 22.5MB.

2. Copy the files—perhaps using some type of compression, given the file sizes—and protect the media to which you have backed up. This is a copy of your Registry, something that you do not want to fall into the wrong hands!

Backing up the Registry with **NTBACKUP**

Backing up the Windows 2000 Registry using **NTBACKUP** involves backing up the system state data. This is simple, thanks to the improved **NTBACKUP** interface:

1. Select Start To Run, type **NTBACKUP**, and click on OK.

2. Click on the Backup tab.

3. In the section labeled, "Click On To Select The Check Box For Any Drive, Folder, Or File That You Want To Back Up," click on the box next to System State (see Figure 3.3). This backs up the system state data.

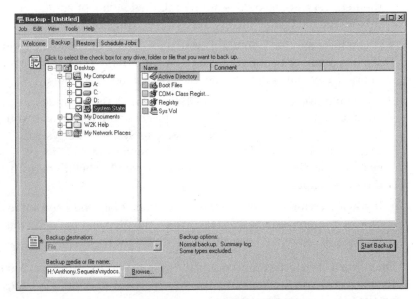

Figure 3.3   Backing up the Registry with **NTBACKUP.EXE**.

Restoring the Registry with **NTBACKUP**

Restoring the Registry using **NTBACKUP** is just as simple as backing up:

1. Select Start to Run, type **NTBACKUP**, and click on OK.

2. Click on the Restore tab.

3. In the section labeled, "Click On To Select The Check Box For Any Drive, Folder, Or File That You Want To Restore," click on the box next to System State. This restores the system state data.

Using **REGBACK** to Back up a Specific Hive

One of the excellent features of the **REGBACK** command prompt utility is its ability to back up a specific Registry hive. The syntax for **REGBACK** is as follows:

```
regback.output.HiveType.HiveName
```

- **Output**—Allows you to specify where the backed up Registry hive is to be stored.

- **HiveType**—Permits you to specify the machine for **HKEY_LOCAL_MACHINE** or users for **HKEY_USERS**.

- **HiveName**—Allows you to specify a subkey beneath either HKLM or HKU for backup.

Using **REGREST** to Restore a Specific Hive

In order to use **REGREST** to restore a specific hive of the Registry, use the following syntax from a command prompt session:

```
regrest BackupFileName SaveFileName HiveType HiveName
```

- **BackupFileName**—Specifies the file name that was backed up.

- **SaveFileName**—Allows you to specify a file name and location for the copy of the existing hive.

3. Registry Disaster Prevention

- **HiveType**—Permits you to specify machine for HKLM or users for HKU.

- **HiveName**—Allows you to specify a subkey beneath either HKLM or HKU for backup.

Saving Registry Keys with Registry Editor

There may be times when you want to create a file that holds the backup information in a selected hive key or key/subkey directly from the current Registry data. Here's how to do this with the Registry Editor:

1. Launch Regedt32.

2. Select the Window menu option that matches the hive key you want to work under; its child window appears.

3. Use the tree control in the left-hand window to navigate to the desired subkey (or select the hive key itself to save it). Click on the key to select it.

4. Select Registry|Save Key from the menu bar, as shown in Figure 3.4.

5. A dialog box asking for the location and file name appears. Enter the appropriate values and click on OK. The Registry key you selected is written to the file as a text breakdown of all the key's subkeys and value entries.

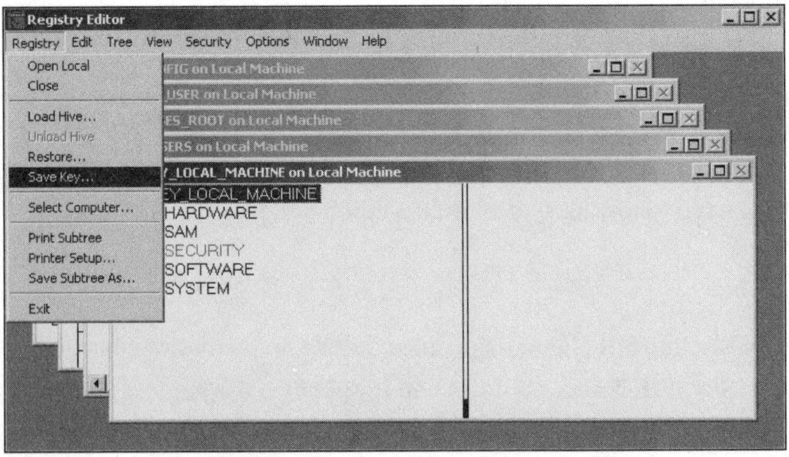

Figure 3.4   The Regedt32.exe Save Key menu option in Windows 2000.

Restoring Registry Keys with Registry Editor

There might be times when you want to restore a selected key, its subkeys, and value entries directly from a previously saved Registry data file. Here's how to do this with the Registry Editor:

1. Launch Regedt32.

2. On the Registry menu, select Restore.

3. A dialog box appears, allowing you to select the file from which to restore. Choose the file and click on OK.

4. If the file is valid, the Registry Editor parses it and finds the Registry key to replace. The Registry Editor then warns that data is about to be overwritten. Click on OK to restore the Registry key from the saved file.

3. Registry Disaster Prevention

System Administration
Tools

In Brief

Windows 2000 has added a new group of system administration tools, in addition to keeping some old ones. The administration tools include the following:

- *Computer Management Console (CMC)*—This Microsoft Management Console (MMC) snap-in provides a single interface for managing many features of the system.

- *Component Services*—Serves the powerful new COM+ system; it allows administrators to configure COM (Component Object Model) server DLLs (dynamic link libraries) into groups for use of advanced services such as Distributed Transactions, Load Balancing, and Queuing.

- *Event Viewer*—Provides lists of system, security, and application events with access to information about interactions between Windows 2000 and its programs, users, and services.

- *System Monitor*—Helps administrators determine where bottlenecks exist on their system by presenting graphs and charts based on feedback from another lower-level utility called the Performance Library.

- *Performance Library*—Makes heavy use of a special type of counter called an *Extensible Counter* and provides information to the System Monitor.

- *Extensible Counter*—Keeps track of non-OS system usage and feeds information to the Performance Library and System Monitor.

Via the Registry, you can affect how these tools work and thus how effectively you can keep track of your Windows 2000 system's performance.

The Computer Management Console

The *Computer Management Console* (an MMC snap-in) is accessed from the Start menu, via Programs|Administrative Tools|Computer Management. Figure 4.1 shows its default user interface.

As you can see in Figure 4.1, the Computer Management Console has an Explorer-style user interface. A tree control on the left gives you access to its various features, each of which displays its data and

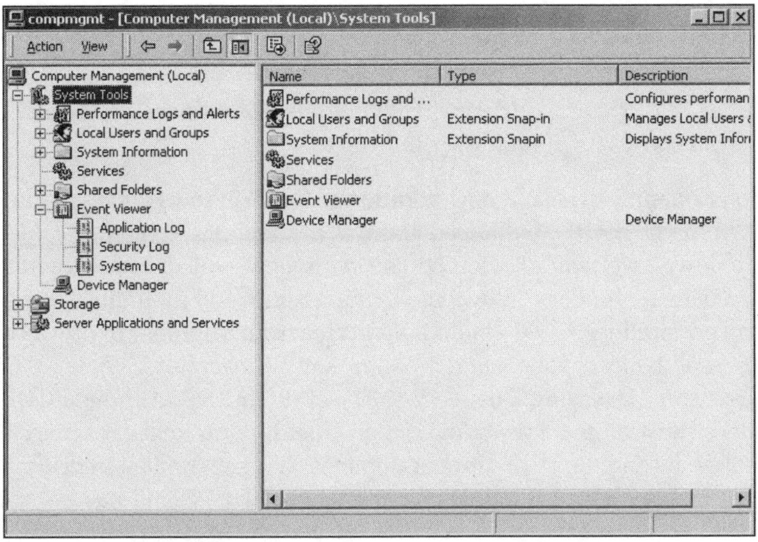

Figure 4.1   The Computer Management Console user interface in
Windows 2000.

specific user controls in the right pane. You can perform the following tasks using the Computer Management Console:

- View performance logs and alerts

- View and modify existing users

- View and modify existing groups

- View system information from the Registry

- Examine services and determine their state and relationships

- View and modify shared folders for the system

- View and modify all the hardware devices currently being used

- Examine the current memory and hard drive storage for the system

- View and administer server applications, such as the Microsoft
  IIS (Internet Information Service)

The Computer Management Console is one of Windows 2000's central
administration utilities and will become a focus for changes and possible problems. In this chapter's "Immediate Solutions" section, there is
a topic that shows administrators how to deal with one of the CMC's
more persistent problems—losing access to custom tools added to the
CMC as snap-ins.

Component Services

Component Services, an MMC snap-in, is accessed from the Start menu, via Programs|Administrative Tools|Component Services. Figure 4.2 shows its default user interface.

Component Services, a new addition to the Windows 2000 system, is used to manage the enhanced features of the COM+ system. COM+ is the newest version of Microsoft's Component Object Model technology. This technology allows applications new and incredible integration possibilities. COM+ includes features from a number of previously released technologies, such as Microsoft Transaction Server (MTS), Microsoft Messaging Queue (MSMQ), and OLE DB. Although COM+ keeps most of its information in a proprietary non-Registry binary database, it (and thus Component Services) has several key Registry entries that are covered in this chapter's "Immediate Solutions" section.

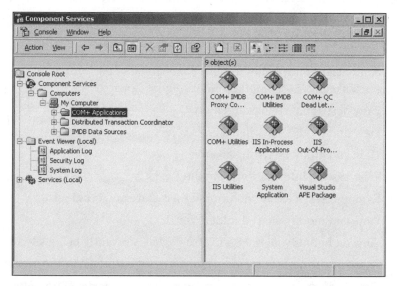

Figure 4.2   The Component Services user interface in Windows 2000.

The Event Viewer

Due to the availability of COM+, Windows 2000 has a much more sophisticated event system than Windows NT 4. This allows administrators a much finer degree of inspection over the behavior of the operating system. The access point for viewing these messages is the Event Viewer MMC snap-in, as shown in Figure 4.3.

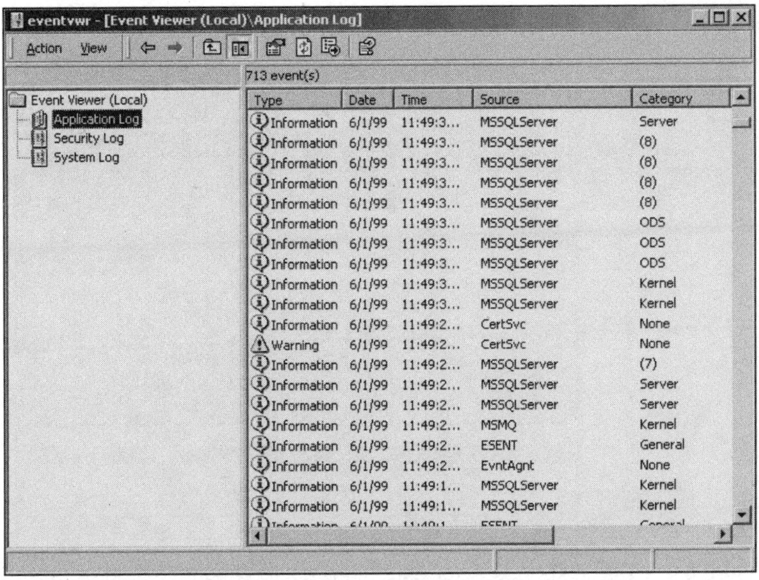

Figure 4.3   The Event Viewer user interface in Windows 2000.

Event Viewer's main job is to allow administrators to watch for specific messages that indicate important or troublesome conditions. Event Viewer provides Source and Event ID information for each event it displays. The Source of the event is one of the most important components because it leads you in the direction of the service or Windows 2000 component that may be troubled or in need of maintenance. The source could be one of several things, such as a service, an application, or a driver. Sometimes the Source name can be somewhat cryptic. Microsoft TechNet can help decipher all of them. For more information about TechNet—see **www.microsoft.com/technet**. Event Viewer has several key Registry entries that are covered in the "Immediate Solutions" section of this chapter.

The Event ID is also very important in researching a problem associated with a particular event. Often searching Microsoft's TechNet database using the Event ID returns the exact cause and solution for the problem you are having.

System Monitor

System Monitor is accessed from the Start menu, using Programs| Administrative Tools|Performance. Once a standalone utility, it is now integrated into the MMC system. Figure 4.4 shows the System Monitor's default user interface.

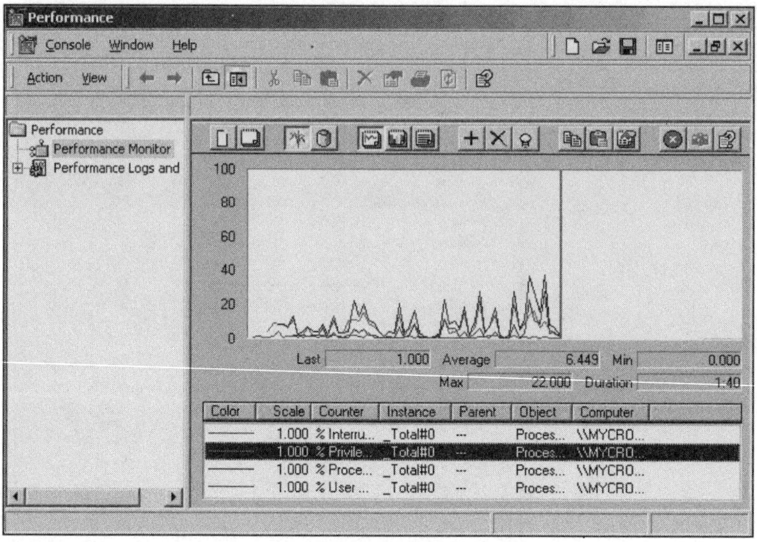

Figure 4.4   The System Monitor user interface in Windows 2000 as viewed in the Performance console.

As you can see in Figure 4.4, System Monitor integrates the standard Windows user-interface features, including a menu bar, status bar, and toolbar. System Monitor allows administrators to perform the following:

- View a chart
- View existing alerts
- View the output log file's status
- View the data in a report
- Add a new counter to acquire data
- Modify an existing counter
- Update the data in a counter
- Add a bookmark to a selected log file along with a comment
- Display System Monitor's options

System Monitor is a complex and powerful application. All Windows 2000 administrators should take the time to become thoroughly familiar with it.

The Performance Library

Underlying the user-friendly interface of the System Monitor is the Windows 2000 Performance Library—the utility that actually collects the data. The Performance Library can compile information on the following system elements:

- Browser
- Cache
- Logical disk
- Memory
- NBT connection
- NetBEUI (NetBIOS Enhanced User Interface)
- NetBEUI resource
- NWLink (NetWare Link) IPX (Internetwork Packet Exchange)
- NWLink NetBIOS (Network Basic Input/Output System)
- NWLink SPX (Sequenced Packet Exchange)
- Application objects
- Paging file
- Physical disk
- Processes
- CPU
- RAS (Remote Access Service) port
- RAS total
- Redirector
- Server
- Server work queues
- System
- Telephony
- Threads

Extensible Counters

All the information recorded by System Monitor uses counters. *Counters* are system objects that store information in a numerical format. There are two counter types: standard system counters and extensible counters. Extensible counters are used for most of the information you'll want to collect with System Monitor because they are used for all the data from non-OS sources.

Immediate Solutions

Hiding the Administration Tools

On some systems, you might not want the Administration Tools to appear on the Start menu. Why have certain users nosing around Computer Management, Event Viewer, and the like? Here is the easy Registry method for hiding the display of these tools:

1. Launch Regedt32.

2. Select the Window menu option for **HKEY_CURRENT_USER**.

3. Use the tree control in the left-hand window to navigate to the **Software\Microsoft\Windows\CurrentVersion\Explorer\Advanced** subkey. Double-click on the subkey to expand it.

4. Double-click on the **StartMenuAdminTools** entry. A value of Yes means that the administrative tools will display in the Start menu. A value of No means they will not.

Checking the Availability of Microsoft Management Console Snap-Ins

Microsoft provides a substantial set of MMC snap-ins, and vendors who provide service applications for Windows 2000 usually provide snap-ins as well. A common problem that arises from this situation is that an invalid Registry entry results in problems accessing the MMC snap-in. The following steps show you how to deal with this situation:

1. Launch Regedt32.

2. Select the Window menu option for **HKEY_LOCAL_MACHINE**.

3. Use the tree control in the left-hand window to navigate to the **SOFTWARE\Microsoft\MMC\SnapIns** subkey. Double-click on the subkey to expand it.

4. A group of subkeys exists below the SnapIns subkey. The subkeys are COM CLSIDs (Class IDs, which are unique text names) values. If you have the CLSID of the missing or misbe-

having snap-in, check to determine whether the CLSID is in the listed subkeys. Otherwise, examine each one for its **NameString** value until you find the right one. If you don't find the entry, you need to reinstall the software (or at least its Administration tools). If you find the entry, move to the **HKEY_CLASSES_ROOT** window, search for its CLSID value, and check its **InProcServer** value entry to make sure the snap-in COM DLL is at the location given. If not, either move the snap-in to the specified location or change the entry to the location of the DLL.

Fixing Lost Component Services Metadata DLLs

Component Services is the name for the new COM+ management system in Windows 2000. COM+ provides COM servers with many new and powerful features, but it's essentially built on top of the existing COM arrangement, which in turn depends heavily on the Registry. COM+ can fail if one or two critical COM Registry entries are damaged or incorrect. Here is where to check to fix such a potentially large problem:

1. Launch Regedt32.

2. Select the Window menu option for **HKEY_CLASSES_ROOT**.

3. Use the tree control in the left-hand window to navigate to the **ComPlusMetaDataServices** subkeys. Double click on the subkeys to expand them.

4. Write down the CLSID subkey value (it has no name). Search HKCR for it. When you find the entry, check to ensure that the DLL for the COM sever is in the indicated location. If not, you need to reinstall COM+2 or determine where the DLL is and either move it or change the Registry entry.

TIP: There is a group of COM+ entries. You might need to check all of them using the technique described in this "Immediate Solution" until you locate the one that has been corrupted.

Related solution:	Found on page:
Saving Registry Keys with Registry Editor	66

4. System Administration Tools

Modifying Event Viewer's Low Disk Space Warning

Windows 2000 has Event Viewer log a warning whenever a disk volume becomes 90 percent full. This is a nice feature for ensuring that we do not have system performance issues due to a lack of disk space. Wouldn't it be nice if we could control the threshold value that triggers the warning? After all, 90 percent might not work for all us. You can change this value by means of this Registry modification:

1. Launch Regedt32.

2. Select the Window menu option for **HKEY_LOCAL_MACHINE**.

3. Use the tree control in the left-hand window to navigate to the **SYSTEM\CurrentControlSet\Services\LanmanServer\Parameters** subkey. Double-click on the subkey to expand it.

4. From the Edit menu, choose **Add** value.

5. Enter **DiskSpaceThreshold** in the Value Name field.

6. Change the Data Type to REG_DWORD.

7. Enter the desired percentage value in the Data field.

8. Shut down and restart Windows 2000.

Determining Why Events Are Not Being Logged

Although it is not advertised in the documentation, the powerful new Event system of Windows 2000 is based on the COM+ technology. This in turn means that it depends on a set of key Registry entries so that its COM servers can be located and started when Windows 2000 boots up. If Windows 2000 Events suddenly stop being correctly obtained, a safe bet is that one of the COM server entries in the Registry is corrupted. To check your COM server entries:

1. Launch Regedt32.

2. Select the Window menu option for **HKEY_LOCAL_MACHINE**.

3. Use the tree control in the left-hand window to navigate to the **SOFTWARE\Microsoft\EventSystem\...\EventClasses** subkey. Double-click on the subkey to expand it.

4. Click on each of the subkeys below the EventClasses key. Check each description until you find the one for the event you are missing. Then, search HKCR for its CSLID (the value of the subkey). Ensure the servicing DLL for the COM server is in the proper location. If not, change the Registry entry or move the DLL.

Changing the Default Event Logging Setting of the Windows 2000 Performance Library

In addition to using System Monitor, you can keep track of system problems via the Event Viewer. You can alter the Registry to control how the Performance Library (System Monitor's workhorse code) sends events to the Event Log that the Event Viewer uses:

1. Launch Regedt32.
2. Select the Window menu option for **HKEY_LOCAL_MACHINE**.
3. Use the tree control in the left-hand window to navigate to the **SOFTWARE\Microsoft\WindowsNT\CurrentVersion\Perflib** subkey. Click on the subkey to select it.
4. Double-click on the **EventLogLevel** value to open it in a DWORD Editor. If this value does not exist already, you can create it. EventLogLevel can be set to one of the four values shown in Table 4.1.

WARNING! This event logging Registry entry affects only Performance Library, not System Monitor. The Immediate Solution "Determining Why Events Are Not Being Logged" earlier in this chapter affects the display System Monitor utility, which must report its own errors.

Table 4.1 **EventLogLevel** values for the Performance Library utility.

Value	Effects
0	No events from Performance Library are logged
1	Only errors are logged; this is the default setting (Event Codes 1000-1013)
2	Errors and warnings are logged (Event Codes 1000-2002)
3	Errors, warnings, information, and success/failures are logged (Event Codes 1000-3000)

(right margin, vertical text) 4. System Administration Tools

Reducing the Buffer Testing Level by the Windows 2000 Performance Library for Extensible Counters

Performance Library uses extensible counters to actually acquire information from nonsystem applications. Extensible counters contain a memory buffer and because the OS does not allocate this buffer, Windows 2000 normally tests it exhaustively before accepting it. In reality, this needs to be done only for new applications whose integrity is in doubt. After the buffers have been shown to be reliable, the testing is unnecessary overhead. To modify the Registry so as to reduce buffer testing overhead for extensible counters:

1. Launch Regedt32.

2. Select the Window menu option for **HKEY_LOCAL_MACHINE**.

3. Use the tree control in the left-hand window to navigate to the **SOFTWARE\Microsoft\WindowsNT\CurrentVersion\Perflib** subkey. Double-click on the subkey to select it.

4. If you do not find an **ExtCounterTestLevel** value under this key, use the Edit|Add Value menu option to create one, with a type of DWORD. Otherwise, use the DWORD Editor to change the ExtCounterTestLevel value to one of the settings shown in Table 4.2.

Table 4.2   ExtCounterTestLevel values for the Performance Library utility.

Value	Effects
1	High testing level; exhaustive testing of all buffers and pointers; default
2	Medium testing; only pointer validity and buffer lengths are checked
3	No testing done

Preventing Timeouts of Extensible Counters by the Windows 2000 Performance Library

One common problem you might encounter with System Monitor is a graph or chart that shows no data, even though you know a process is taking place. This can happen for several reasons, but the most common is that the extensible counters being used to keep track of the information are timing out due to a delay in the process starting up. You can correct this problem by changing the Registry as follows:

1. Launch Regedt32.

2. Select the Window menu option for **HKEY_LOCAL_MACHINE**.

3. Use the tree control in the left-hand window to navigate to the **SOFTWARE\Microsoft\WindowsNT\CurrentVersion\Perflib** subkey. Double-click on the subkey to expand it.

4. If you do not find an **OpenProcedureWaitTime** value under this key, use the Edit|Add Value menu option to create one, with a type of DWORD. Otherwise, double-click on the OpenProcedureWaitTime value to open it in the DWORD Editor. The default value is 5,000 msec., or five seconds. Increase the value to a more reasonable figure and try System Monitor again until you obtain valid data—bearing in mind that too large a value can negatively impact system performance.

Using the Collect Timeout Value

New to Windows 2000 is a collect timeout value that the Performance Library can use. If the value is present, **perflib** sets up a timeout procedure internally. If a system monitor extension DLL's **Collect** function does not return within the time specified (in milliseconds) in this Registry value, an event (1015) is posted to the Event Log. To edit this value, follow these steps:

1. Launch Regedt32.

2. Select the Window menu option for **HKEY_LOCAL_MACHINE**.

3. Use the tree control in the left-hand window to navigate to the **SYSTEM\CurrentControlSet\Services\(*servicename*)\ Performance** subkey.

4. Use the **collect timeout** value to set the timeout procedure.

Changing the _Total Instance Name in System Monitor

Starting with Windows NT 4, Performance Monitor displays a _Total Instance for each of several categories, summarizing their data. You can alter this display name for easier readability by changing the Registry as follows:

1. Launch Regedt32.

2. Select the Window menu option for **HKEY_LOCAL_MACHINE**.

3. Use the tree control in the left-hand window to navigate to the **SOFTWARE\Microsoft\WindowsNT\CurrentVersion\Perflib** subkey. Double-click on the subkey to expand it.

4. Double-click on the **TotalInstanceName** value to open it in a String Editor. Enter a value you find easier to locate or that makes better sense for your installation.

Enabling the Use of Unicode Process Header Names in System Performance Monitor

Windows 2000 supports the use of Unicode characters, but this incurs a performance hit in some applications. For this reason, it is normally disabled, allowing only ANSI (American National Standards Institute) character usage. System Monitor is one place in which this feature is turned off by default. You can change the Registry so that Unicode support is enabled for System Monitor by following these steps:

1. Launch Regedt32.

2. Select the Window menu option for **HKEY_LOCAL_MACHINE**.

3. Use the tree control in the left-hand window to navigate to the **SOFTWARE\Microsoft\WindowsNT\CurrentVersion\Perflib** subkey. Double-click on it to expand it.

4. Double-click on the **CollectUnicodeProcessNames** value to open it in the DWORD Editor. You might have to add this value to the Registry if it does not exist. The default value is 0, which disables the feature. Set the CollectUnicodeProcessNames to 1 to allow using Unicode strings in process headers with System Monitor.

Changing the Paged Memory Pool Size

One factor that can drastically affect how Windows 2000 performs is the size of the paged and nonpaged memory pools. You can set the paged pool advisory values (which the operating system uses in its determination of how memory is actually managed) directly in the Registry by following these steps:

1. Launch Regedt32.

2. Select the Window menu option for **HKEY_LOCAL_MACHINE**.

3. Use the tree control in the left-hand window to navigate to the **SYSTEM\CurrentControlSet\Control\Session Manager\Memory Management** subkey. Double-click on the subkey to expand it.

4. Double-click on the **PagedPoolSize** value to open it in the DWORD Editor. Set a value that experience suggests is more effective, given the demands on your system. Exit Registry Editor and restart the system so the value takes effect.

Related solution:	Found on page:
Enabling Paged Memory Quotas	168

Changing the Nonpaged Memory Pool Size

You can set the nonpaged pool advisory values (which the operating system uses in its determination of how memory is actually managed) directly in the Registry by following these steps:

1. Launch Regedt32.

2. Select the Window menu option for **HKEY_LOCAL_MACHINE**.

3. Use the tree control in the left-hand window to navigate to the **SYSTEM\CurrentControlSet\Control\Session Manager\Memory Management** subkey. Double-click on the subkey to expand it.

4. Double-click on the **NonPagedPoolSize** value to open it in the DWORD Editor. Set a value that experience suggests is more effective given the demands on your system. Exit Registry Editor and restart the system to have the value take effect.

4. System Administration Tools

Related solution:	Found on page:
Enabling Nonpaged Memory Quotas	168

Forcing a Windows 2000 Crash

Want to test how your system will respond to a fatal error (better known as a Blue Screen of Death)? You can actually force Windows 2000 to crash using this Registry modification:

1. Launch Regedt32.

2. Select the Window menu option for **HKEY_LOCAL_MACHINE**.

3. Use the tree control in the left-hand window to navigate to the **System\CurrentControlSet\Services\i8042prt\Parameters** subkey. Double-click on the subkey to expand it.

4. From the Edit menu, choose **Add** value.

5. Enter **CrashOnCtrlScroll** in the Value Name field.

6. Change the Data Type to REG_DWORD.

7. Enter 1 for the value in the Data field.

8. Shut down and restart Windows 2000. To force a crash hold down the Ctrl key on the right-hand side and press the Scroll Lock key twice.

Checking How Long Windows Has Been "Up"

You might notice that not one of the Administration tools explored in this chapter provides a simple and immediately accessible display of how long a particular Windows 2000 system has been functional since the last reboot. Here is a simple way to acquire this information:

1. Launch the Command Prompt.

2. Type the command:

```
net statistics workstation|more
```

3. The second line of information displayed states "Statistics since...*date time*". This is how long your workstation service has been functional and should be extremely close to the total time your system has been operational since last boot.

Chapter 5

Windows 2000 User Interfaces

In Brief

Windows 2000 comes with three *shells*, or user interfaces: the Win32 Console (or command prompt shell), the Explorer shell (identical to Windows 95/98), and the Program Manager shell (very much like Windows 3.x). Each shell has its proponents and detractors in the NT/Windows 2000 community, and you can make each more effective by tweaking some little-known Registry entries.

Win32 Console Shell

The *Win32 Console shell*, also called the command prompt shell, is essentially a visual wrapper around a Windows 2000 version of the DOS command prompt. Of course, you are running Windows 2000 instead of DOS, but the "look and feel" is very Unix/DOS-like. There's a good reason for this: The progenitor of Windows 2000 (Windows NT) was actually a recreation of AT&T Unix System 5/7. Figure 5.1 shows the Console shell in operation.

You can perform many essential Windows 2000 tasks (such as navigating hard drives, obtaining network information, and so on) from the command prompt shell without invoking another window. The Registry contains settings that control how these windows are displayed, as well as a number of other Console shell features.

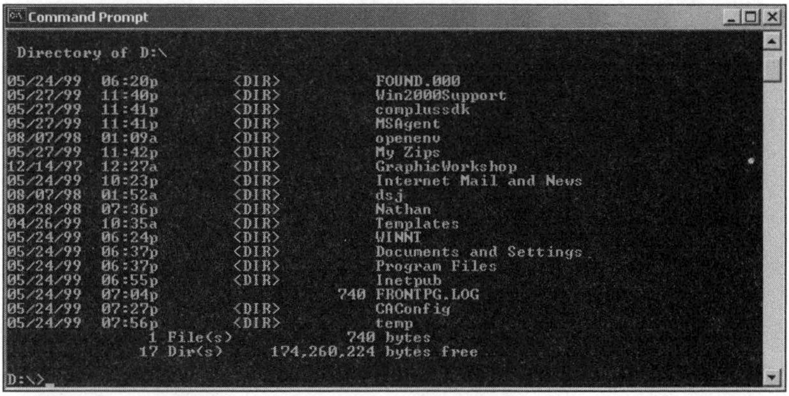

Figure 5.1   The Console shell in Windows 2000.

5. Windows 2000 User Interfaces

TIP: *You used to be able to configure many of the Console shell's settings from the Control Panel's Console applet, as well as directly within the Registry. In Windows 2000, the applet is no longer available.*

Explorer Shell

Perhaps the most exciting feature about NT 4 was the addition of the *Explorer shell,* which is the same user interface as that for Windows 95 and now Windows 98. Windows 2000 expands the features of the Explorer shell and makes it the primary shell for the operating system. Figure 5.2 shows the Explorer shell in operation.

The Explorer shell is made up of the following elements, most of which can be controlled via the Registry:

- Network Neighborhood
- My Briefcase
- Recycle Bin
- My Computer
- Printers
- Dial-Up Networking (DUN)/TAPI (Telephony Application Programming Interface)
- Control Panels

The Explorer shell also provides features such as Autorun for CDs and desktop shortcuts, also called *links.* These features use the Reg-

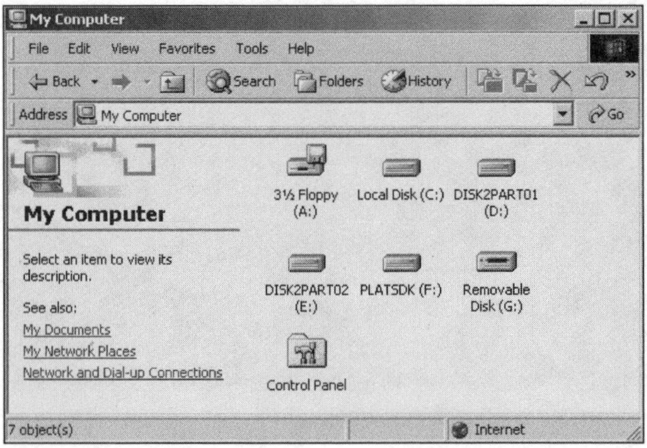

Figure 5.2   The Explorer shell in Windows 2000.

istry to associate themselves with applications and COM Servers via file extensions; these combinations of linked files and associated applications are called *shell objects*. Each shell object has a *context menu* (also called a *shortcut menu*) associated with it and controlled by Registry entries.

Program Manager Shell

Prior to the addition of the Explorer shell, NT used the Windows 3.x shell, which is now called the *Program Manager shell*. Figure 5.3 shows the Program Manager shell in operation in Windows 2000.

You can customize each user's access to Program Manager via the Registry to control the following features:

- Edit level (from no restriction to a complete lockout)
- Exit settings (the **Exit** command can be removed to prevent closing)
- File usage (the File menu can be removed)
- Program usage (the **Run** command can be removed)
- Settings (saved settings can be disabled on exit)
- Program groups (program groups can be hidden)

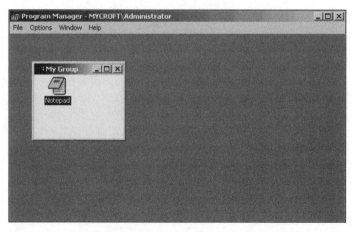

Figure 5.3   The Program Manager shell in Windows 2000.

Tweak UI 1.33

Perhaps the safest and easiest way to make many of the Registry changes mentioned in this chapter is to use this nifty tool. Tweak UI provides a GUI interface for making Registry changes that affect the Explorer interface. As of this writing, you can download the tool from Microsoft at: **www.microsoft.com/ntworkstation/downloads/ PowerToys/Networking/NTTweakUI.asp**.

TIP: Even though Microsoft makes this tool available for download, do not try to get support from them for this tool. It simply will not happen. This is a use-at-your-own-risk deal.

Immediate Solutions

Enabling File Name Completion for the Command Prompt Shell

The Unix C shell has a nifty feature called *file name completion* that takes an incomplete pathname in a command line and, when the Tab key is pressed, searches all available paths to find the closest match. By changing a Registry entry, you can enable this feature for the Windows 2000 command prompt shell as well:

1. Launch Regedt32.

2. Select the Window menu option for **HKEY_CURRENT_USER**.

3. Use the tree control in the left-hand window to navigate to the Software\Microsoft\Command Processor subkey. Double-click on the subkey to expand it.

4. Locate the **CompletionChar** entry. If **CompletionChar** is not present, create it with a type of REG_DWORD. Either way, enter an ASCII (American Standard Code for Information Interchange) value (the default is 9 for the Tab key, but you can use any key you like as long as you don't need it for actual file names) to trigger file completion.

TIP: For an example of how powerful this feature is, try this: type **cd \d**, then press Tab repeatedly as you start cycling through the directories on your hard drive starting with those beginning with the letter d. Think that's cool? Try Shift+Tab to cycle backwards!

Changing Default Options for the Command Prompt Shell

The command prompt shell has a surprising number of configurable options that can be accessed via the Registry. To view the configurable options, follow these steps:

1. Use Registry Editor to access the Windows Registry.

2. Select the Window menu option for **HKEY_CURRENT_USER**.

Table 5.1   Setting the Console shell's default properties in the Registry.

Value Name	Type	Settings' Effects
CursorSize	REG_DWORD	Percentage of a char for the cursor (in hex)
FullScreen	REG_DWORD	1 for full screen; 0 for windowed (Intel only)
FaceName	REG_SZ	Name of font for the console (must be local)
FontFamily	REG_DWORD	0 for TrueType, or other value if not
FontSize	REG_DWORD	32 bits for width (low 16) and height (high 16)
FontWeight	REG_DWORD	0 for normal, or other for bold, italic, and so forth
HistoryBufferSize	REG_DWORD	Hex value of recallable commands
InsertMode	REG_DWORD	0 for insert mode; 1 for overwrite mode
NumberOfHistoryBuffers	REG_DWORD	Hex value (4 is default) of history memory buffs
PopupColors	REG_DWORD	8 bits for foreground menu color (low 4) and background (high 4)
QuickEdit	REG_DWORD	0= no mouse edits; 1= mouse edits
ScreenBufferSize	REG_DWORD	32 bits for screen buffer, width= low 16
ScreenColors	REG_DWORD	8 bits for foreground screen color (low 4) and background (high 4)
WindowSize	REG_DWORD	32 bits for console windows size (width low)

3.   Use the tree control in the left-hand window to navigate to the Console subkey. Click on the subkey to select it. Table 5.1 lists a number of useful value entries under the Console key, along with the values' data types and the effects of the values' settings.

TIP: You can set some of these options with a GUI inside Windows 2000. From a command prompt session, click on the control box in the upper-left corner of the command prompt window and choose Properties from the menu.

Setting Console Window Options for Specific Windows in the Command Prompt Shell

If you make any changes to the appearance of a console window in the command prompt shell in Windows 2000, the settings you change are saved in the Registry under the name of that application to be reused the next time it is opened. Conversely, you can create default settings this way without opening the application first. To set console window options using the Registry Editor, follow these steps:

1. Launch Regedt32.

2. Select the Window menu option for **HKEY_CURRENT_USER**. Its child window appears. Maximize the window for easier use.

3. Use the tree control in the left-hand window to navigate to the Console\SubKeyNames subkey. If this subkey is present, click on it to select it. Otherwise, create the subkey using Edit|Add Key, then click on it to select it.

4. Create a new subkey with the same name as the desired application. Under the new subkey, enter a value name for any of the properties shown previously in Table 4.1, and set the desired value. Exit Regedt32, and, whenever you start the named application, it will automatically take on the entered settings.

Displaying Your Own "Tips of the Day" at Logon

If you are using Windows 2000 Professional you can display your own tips at logon—which can really be fun and educational. If you are using some version of Windows 2000 Server, you can have a special Configure The Server application display at logon using the following Registry tweaks. All might come in really handy:

1. Launch Regedt32.

2. Select the Window menu option for **HKEY_LOCAL_MACHINE**.

3. Locate the **Software\Microsoft\Windows\CurrentVersion\ Explorer\Tips** subkey. Double-click on the subkey to display its values in the right-hand window.

4. Notice that the tip values are stored as sequentially named REG_SZ values.

5. Replace any of the tips with your own tips by double-clicking on the values to edit them.

6. Select the Window menu option for **HKEY_CURRENT_USER**.

7. Locate the **Software\Microsoft\Windows\CurrentVersion\ Explorer\Tips\Show** subkey.

5. Windows 2000 User Interfaces

8. This entry controls whether or not the tip or configuration applet in Server is displayed. A value of 0 disables the tip display, while a value of 1 enables it.

9. Locate the **Software\Microsoft\Windows\CurrentVersion\ Explorer\Tips\Next** subkey. This controls which tip appears next. Use the numbered tips to control which tip you want to display first—Windows 2000 automatically increments the value as each tip is displayed.

TIP: *If any of these values do not exist already in the Registry—you can add them. You add them with the Edit\Add Value menu option.*

Disabling "Extra" Windows Animations

Windows 2000 carries on the tradition of earlier Explorer interfaces and adds "extra" animations to show windows actually being minimized to and maximized from the taskbar. This extra animation was designed to help new users to the interface actually see where their windows are going when minimized. This might be great for a beginner, but most of us cannot stand this slowdown and extra little hit on resources. Let's speed things up by eliminating these additional effects:

1. Launch Regedt32.

2. Select the Window menu option for **HKEY_CURRENT_USER**.

3. Use the tree control in the left-hand window to navigate to the **Control Panel\Desktop\WindowsMetrics** subkey. Double-click on the subkey to expand it.

4. Create a new value with the name **MinAnimate** of type REG_SZ.

5. Set the value to 0.

Enabling Separate Process Creation for Each Explorer Instance in the Explorer Shell

A potentially nasty behavior of the Explorer shell is that all copies of the Explorer run in one process, on different threads. This can wreak havoc if one Explorer experiences a problem, as you might imagine. If a computer has at least 24MB of physical RAM and a 100MHz processor (or faster), you can tweak the Registry to avoid this small disaster:

1. Launch Regedt32.

2. Select the Window menu option for **HKEY_CURRENT_USER**.

3. Use the tree control in the left-hand window to navigate to the **Software\Microsoft\Windows\CurrentVersion\Explorer** subkey. Double-click on the subkey to expand it.

4. Create a new value with the name **DesktopProcess**, of type REG_DWORD. Set its value to 1. Exit Registry Editor and reboot to activate the new arrangement.

Preventing Restoring Open Explorer Windows on Restart in the Explorer Shell

One aspect of the Explorer shell sometimes necessitates extra work for the user: When you close Windows 2000 and restart it, any open Explorer windows are automatically restored. If a user wants this behavior, it is quite nice. But some users do not and they must then manually close the open Explorer windows. Fortunately, you can change a Registry entry to deactivate this default action:

1. Launch Regedt32.

2. Select the Window menu option for **HKEY_CURRENT_USER**.

3. Use the tree control in the left-hand window to navigate to the **Software\Microsoft\Windows\CurrentVersion\Policies\Explorer** subkey. Double-click on the subkey to expand it.

4. Create a new value with the name of **NoSaveSettings** if one does not already exist, of type REG_BINARY. Set the value to hexadecimal 1. Exit Registry Editor, close all applications, log out, and then log back on. Making this change has the drawback of not allowing you to resize the taskbar or change the position of shortcuts after you place them on the desktop for the first time.

5. Windows 2000 User Interfaces

Forcing Automatic Reboot on Crashing in the Explorer Shell

If the OS goes down, Windows 2000 is set to automatically reboot. However, if the Explorer shell crashes (typically this is evidenced with the desktop showing no icons or taskbar), you might have to use the keyboard shortcuts to log off and log on to restart the shell. To avoid this hassle, make the following Registry tweak:

1. Launch Regedt32.

2. Select the Window menu option for **HKEY_LOCAL_MACHINE**.

3. Use the tree control in the left-hand window to navigate to the **SOFTWARE\Microsoft\WindowsNT\CurrentVersion\Winlogon** subkey. Double-click on the subkey to expand it.

4. Look for the **AutoRestartShell** value. If AutoRestartShell has a value of 0, edit it with the DWORD Editor, set it to 1, and reboot. The next time the Explorer shell crashes, the computer will automatically restart the shell without having to log off and on again—a real timesaver!

TIP: If AutoRestartShell does not exist, add it with the Edit|Add Value menu option.

Disabling the File Menu in the Windows Explorer

The File menu in the Windows Explorer has many powerful menu items, including Rename and Delete to name a few. You can remove the File menu completely from the Windows Explorer with this handy Registry modification:

1. Launch Regedt32.

2. Select the Window menu option for **HKEY_CURRENT_USER**.

3. Navigate to the **Software\Microsoft\Windows\ CurrentVersion\Policies\Explorer** subkey and double-click on it to expand it.

4. Add the **NoFileMenu** REG_DWORD value and set it to 1.

Disabling File Type Autochecking in the Program Manager Shell

If you are using the Program Manager shell over a network running Windows 2000, you might notice considerable network traffic being generated by the Program Manager itself. This is due to a feature called *file type autochecking*. File type autochecking takes a file name from the Run, New Item, or Edit Item dialog box and performs a network call to determine if the file is an executable file. If so, a Run In Separate Memory Space checkbox is displayed; otherwise, it is not. You can remove this load on your system with the following Registry edit:

1. Launch Regedt32.

2. Select the Window menu option for **HKEY_CURRENT_USER**.

3. Locate the **Software\Microsoft\WindowsNT\Current Version\Program Manager\Settings** subkey. Double-click on the subkey to display its values in the right-hand window.

4. You can take one of two actions: set the **CheckBinaryTimeout** REG_DWORD value to a higher number (500 is the default; a larger value reduces the number of times the remote check is made because it is based on the user's typing speed) or set the CheckBinaryType REG_DWORD to 0 to turn off the checking completely.

Setting Program Group Display Order in the Program Manager Shell

There is a set order in how open Program Manager windows are displayed when the application is started. If you are really picky about your user interface you can change this order via the Registry in two ways:

1. Launch Regedt32.

2. Select the Window menu option for **HKEY_CURRENT_USER**.

3. Locate the **Software\Microsoft\WindowsNT\Current Version\ Program Manager\Settings** subkey. Double-click on the subkey to display its values in the right-hand window.

4. At this point, you can take one of two actions: Set the **Order** REG_SZ value to a list of window names in the reverse order

you want them to display (last to first) or set the
UnicodeOrder REG_SZ value to the list of window names in
the order you want them to be shown (first to last).

TIP: You might need to add these values if they do not already exist.

Adding a Custom Shell for All Users

Custom shells are available for Windows 2000, but to get them to work,
you have to edit the Registry. To use the Registry to add a custom
shell, follow these steps:

1. Launch Regedt32.

2. Select the Window menu option for **HKEY_LOCAL_MACHINE**.

3. Locate the **SOFTWARE\Microsoft\WindowsNT\Current
   Version\ Winlogon** subkey. Double-click on the subkey to
   display its values in the right-hand window.

4. In the **Shell** value, replace the displayed file name and
   pathname with the custom file name and pathname. Make sure
   the custom shell is well tested and documented, unless you like
   reconstructing your system from scratch.

TIP: You might need to add the Shell value before setting it.

Displaying Windows 2000 Version Information

Would you find it useful to display very specific version information
about your Windows 2000 computer on the desktop? If so, you should
make this little Registry change:

1. Launch Regedt32.

2. Select the Window menu option for **HKEY_CURRENT_USER**.

3. Locate the **Control Panel\Desktop** subkey. Double-click on
   the subkey to display its values in the right-hand window.

4. Edit the **PaintDesktopVersion** value so the value is set to 1.

TIP: Look for the version information in the lower-right corner.

Eliminating the Documents Content from the Start Menu

Do you view the Documents option of the Start menu as a bit of a security risk? After all, this area shows the last 15 documents with which you have worked on the system. If you would like to disable this menu option, use this handy Registry modification:

1. Launch Regedt32.

2. Select the Window menu option for **HKEY_CURRENT_USER**.

3. Locate the **\Software\Microsoft\Windows\CurrentVersion\ Policies\Explorer** subkey. Double-click on the subkey to display its values in the right-hand window.

4. Add the value **NoRecentDocsMenu** as type REG_DWORD and set it to 1.

Speeding up the Menus

Some users are faster than others. For the fast crowd, the time it takes a submenu to appear when the mouse moves over it is measured in geological epochs. Fortunately, you can use the following steps to tweak the Registry and allow users to have their own settings for this action:

1. Launch Regedt32.

2. Select the Window menu option for **HKEY_CURRENT_USER**.

3. Locate the **Control Panel\Desktop** subkey. Double-click on the subkey to display its values in the right-hand window.

4. Edit the **MenuShowDelay** value, with the DWORD Editor. This value represents the time in milliseconds from when the mouse moves over a menu item with a submenu and when the submenu appears.

Setting Wallpaper

You might want to customize the wallpaper for users. Here's how to set wallpaper via the Registry:

1. Launch Regedt32.

2. Select the Window menu option for **HKEY_CURRENT_USER**.

5. Windows 2000 User Interfaces

3. Locate the **Control Panel\Desktop** subkey. Double-click on the subkey to display its values in the right-hand window.

4. Edit the **Wallpaper** value with the String Editor. This value is the path-name and file name of the wallpaper graphic.

Setting Screen Background Color

If you don't use wallpaper, you can customize the background color for the user via the Registry by following these steps:

1. Launch Regedt32.

2. Select the Window menu option for **HKEY_CURRENT_USER**.

3. Locate the **Control Panel\Desktop** subkey. Double-click on the subkey to display its values in the right-hand window.

4. Edit the **Background** value with the REG_DWORD Editor. The Background value indicates the color of the background screen (in RGB order) if a wallpaper graphic is not selected.

TIP: You might need to add the Background value before setting it.

Get Detailed Logon and Logoff Information

Detailed information is excellent for troubleshooting logon and logoff problems, or for those who are really curious about just what Windows 2000 is doing in all that booting or shutting down time. You can have Windows 2000 display all the intimate details during these periods with a simple Registry change:

1. Launch Regedt32.

2. Select the Window menu option for **HKEY_LOCAL_MACHINE**.

3. Navigate to the **SOFTWARE\Microsoft\Windows\Current Version\Policies\System** subkey. Double-click on the subkey to display its values in the right-hand window.

4. On the Edit menu, choose Add Value name **VerboseStatus**, as a REG_DWORD data type. A data value of 0, the default, causes the system to display normal status messages. A data value of 1 enables verbose status messages in each step of the process of starting, shutting down, logging on, and logging off.

Suppressing Logon Error Messages

You can force Windows 2000 to suppress the display of error messages during logon by manipulating the Registry. These errors are still logged to Event Viewer, but they no longer rudely interrupt the logon process:

1. Launch Regedt32.

2. Select the Window menu option for **HKEY_LOCAL_MACHINE**.

3. Locate the **SOFTWARE\Microsoft\Windows NT\CurrentVersion\Windows** subkey. Double-click on the subkey to display its values in the right-hand window.

4. Add the **NoPopUpsOnBoot** value as a REG_DWORD and assign it a value of 1.

5. Add the **ErrorMode** value as a REG_DWORD and assign it a value of 2.

Setting Logon Wallpaper

You can easily set new wallpaper for the logon screen with this simple Registry modification. This "logon" wallpaper differs from the earlier wallpaper setting in that this wallpaper is displayed behind prior to logon. Follow the following steps:

1. Launch Regedt32.

2. Select the Window menu option for **HKEY_USER**.

3. Locate the **DEFAULT\Control Panel\Desktop** subkey. Double-click on the subkey to display its values in the right-hand window.

4. Edit the **Wallpaper** value to provide the full path and name of the graphic.

5. Windows 2000 User Interfaces

Setting a New Logon Screen Saver

Notice that Windows 2000 sets a screen saver to display if your system is sitting at the Ctrl+Alt+Del To Logon screen for too long. This Registry tweak allows you to select the specific screen saver you would like displayed:

1. Launch Regedt32.

2. Select the Window menu option for **HKEY_USER**.

3. Locate the **DEFAULT\Control Panel\Desktop** subkey. Double-click on the subkey to display its values in the right-hand window.

4. Ensure the **ScreenSaveActive** value is set to 1.

5. Double-click on the **SCRNSAVE.EXE** value and specify the full path and name of the screen saver you want to run. If the screen saver is located in %systemroot%\system32, you do not have to worry about specifying the path—just provide the name; for example, ssbezier.scr.

Setting Logon Screen Saver Timeout

A minor, but potentially dangerous, security hole can occur when a user gets called away from a workstation just when he or she starts to log on, but hasn't actually logged on. You can set a delay time via the Registry for the screen saver to come up, which you can configure to require its own password, and thus fill this little hole. To set a logon timeout, follow these steps:

1. Launch Regedt32.

2. Select the Window menu option for **HKEY_USER**.

3. Locate the **DEFAULT\Control Panel\Desktop** subkey. Double-click on the subkey to display its values in the right-hand window.

4. Edit the **ScreenSaveTimeOut** value. This value is the time in seconds before the screen saver appears during the logon process when no user input has occurred. Set ScreenSave TimeOut to a lower value to plug the security hole (but not so low that users cannot log on).

Adding a File Extension to the New Context Menu

You can manually add a file extension to the list displayed in the New Context menu (right-clicking on the desktop) for the Explorer shell. To do so, tweak the Registry as follows:

1. Launch Regedt32.

2. Select the Window menu option for **HKEY_CLASSES_ROOT**.

3. Locate the subkey for the file extension you want to add to the New menu. Double-click on the subkey to display its values in the right-hand window.

4. Add a **ShellNew** subkey to the file extension key. Then, enter a file value of type REG_SZ with the pathname that points to the application that can respond to a DDE (Dynamic Data Exchange) or OLE (Object Linking and Embedding) message to create a new document of the named type.

Turning Off the CD Autorun Feature

The Explorer shell's CD Autorun feature can, at times, present a security hole. It can also be an annoyance when you do not want CD-ROMs automatically executing an applet every time you insert them. Here's how to fix the Registry to disable the Autorun feature:

1. Launch Regedt32.

2. Select the Window menu option for **HKEY_LOCAL_MACHINE**.

3. Locate the **SYSTEM\CurrentControlSet\Services\Cdrom** subkey. Double-click on the subkey to display its values in the right-hand window.

4. Locate the **Autorun** value, of type REG_DWORD. Set Autorun to 0 to disable the CD Autorun feature and set Autorun to 1 to enable it again.

Setting Default Keyboard Behavior

Individual users often require different keyboard settings, depending on their personal typing habits and possible handicaps. Here is how to alter the Registry to enable each user to customize the keyboard:

1. Launch Regedt32.

2. Select the Window menu option for **HKEY_CURRENT_USER**.

3. Locate the **Control Panel\Keyboard** subkey. Double-click on the subkey to display its values in the right-hand window.

4. To change the time interval needed for a key to start repeating, set the **KeyboardDelay** value to a DWORD from 0 through 3, which represent 1/4 of a second through 2 seconds.

5. To change the number of keys inserted during a repeat operation, set the **KeyboardSpeed** value to a number from 0 through 31. 0 indicates two keys inserted per second, and 31 indicates 30 keys inserted per second.

Controlling the State of NumLock at Logon

You can choose whether to have NumLock on or off automatically at startup with this handy little Registry modification:

1. Launch Regedt32.

2. Select the Window menu option for **HKEY_CURRENT_USER**.

3. Locate the **Control Panel\Keyboard** subkey. Double-click on the subkey to display its values in the right-hand window.

4. Edit the **InitialKeyboardIndicators** value and set it to 2 if you would like NumLock automatically turned on at logon. If you want it off set the value to 0.

Setting Default Mouse Behavior

As with the keyboard, you can customize the mouse for each user. Here are the Registry tweaks that will automatically tell the computer which mouse settings to use when the user logs on:

1. Launch Regedt32.

2. Select the Window menu option for **HKEY_CURRENT_USER**.

3. Locate the **Control Panel\Mouse** subkey. Double-click on the subkey to display its values in the right-hand window.

4. To change the time interval necessary for two clicks to be considered a double click, edit the **DoubleClickSpeed** value entry to a DWORD number. The default is on 686. The lower the value, the faster you must click twice to produce a double-click.

5. To enable fast mouse motions, first set the **MouseSpeed** value to 1 or 2 (its default is 0, which signifies no extra speed behavior). Then, enter a value for the **MouseThreshold1** and **MouseThreshold2** values, in pixels. If the mouse moves across MouseThreshold1 pixels in a given time interval, the mouse movement is doubled. If it then moves across MouseThreshold2 pixels in the same interval, the mouse speed is doubled again. This allows very fast mouse motion across a large pixel display.

Enabling Active Window Tracking

Windows 2000 can have the mouse make a window active simply by moving the mouse pointer into that window. This is one change you might want to experiment with to see if you like it. Some do—some do not:

1. Launch Regedt32.

2. Select the Window menu option for **HKEY_CURRENT_USER**.

3. Locate the **Control Panel\Mouse** subkey. Double-click on the subkey to display its values in the right-hand window.

4. Set the **ActiveWindowTracking** value to 1.

Enable Snap to Default Mouse Behavior

This is another change that users either love or hate. You can have the mouse pointer automatically snap to the default button in a window via the Registry with the following tweak:

1. Launch Regedt32.

2. Select the Window menu option for **HKEY_CURRENT_USER**.

3. Locate the **Control Panel\Mouse** subkey. Double-click on the subkey to display its values in the right-hand window.

4. Set the **SnapToDefaultButton** value to 1.

Chapter 6

TCP/IP and the Internet

In Brief

The Windows 2000 implementation of TCP/IP (Transmission Control Protocol Internet Protocol) provides Microsoft Windows networks with the crucial peer-to-peer link for Unix connectivity and beyond. In order to fully understand the networking parameters contained within the Windows 2000 Registry (and appropriate values for those parameters), a good understanding of essential network architecture is needed. An in-depth discussion of network background is beyond the scope of this book; however, some important facts are provided in this chapter to help enhance your effectiveness as a Windows 2000 administrator. I'll cover Microsoft network protocols, TCP/IP architecture, and DHCP (Dynamic Host Configuration Protocol) so you'll have a context for the "Immediate Solutions" section that follows.

MS Windows Network Protocols

There are more than a few protocols available to Microsoft Windows networks, including TCP/IP, NetBEUI (NetBIOS Enhanced User Interface), AppleTalk, IPX/SPX (Internetwork Packet Exchange/Sequenced Packet Exchange), PPP (Point-to-Point Protocol), SMB (Server Message Block), and DECNET (Digital Equipment Corporation Network). The ability for a particular server to communicate using any protocol depends on the protocols that are installed on the system. In fact, special protocol handling between Windows 2000 computers and other servers can enable an on-the-fly conversion from one protocol to another. The important point to remember is that network protocols are layered. TCP/IP represents a prime example of this layering.

TCP/IP

IP (Internet Protocol) is a protocol responsible for addressing and sending datagrams from one computer on a network to another. A separate technology called Transmission Control Protocol (TCP) is layered on top of IP to provide connection-based communications supporting reliable, sequenced data streams. The combination of the two is used to communicate on the Internet and is usually referred to as *TCP/IP*.

Visualize a driver getting into an automobile. The engine (IP) provides the force necessary to drive on roads, but the driver (TCP) uses the controls of the car (gas pedal, brakes, and steering wheel) to get from point A to point B and travel the network of roads (network cables), and stop lights and road signs (routers). Without the engine, the driver can't go anywhere and without the driver, the engine is useless. So it is with the relationship between TCP and IP.

Windows Sockets

Windows Sockets (WinSock) is a Microsoft implementation of a widely regarded networking system originally developed for Unix. Windows Sockets, in most instances, is compatible with Unix and is one form of communication between a Microsoft Windows network and a Unix network. Windows Sockets is layered on several protocols, much like TCP is layered on IP. Windows Sockets can use the TCP/IP protocol or it can use one of several others, including IPX/SPX and AppleTalk.

NetBEUI

NetBEUI is usually a nonroutable protocol. IBM developed NetBEUI in 1985 and subsequently cross-licensed it to several platforms, including Microsoft Windows. Windows 2000 supports NetBEUI to provide backward compatibility with other systems and computers, such as LAN Manager. NetBEUI can be routed if the Windows 2000 system is running on a token ring (TR) network and TR-Source Routing is activated for the wide area network.

Named Pipes

Named Pipes is layered on several protocols, including Sockets, NetBEUI, TCP/IP, and IPX/SPX, among others. A *pipe* is an area reserved in memory for processing information, usually in the form of autonomous transactions. A process that creates a pipe is called a *pipe server*. A process that accesses an already created pipe is called a *pipe client*. Named Pipes are pipes that have a unique name. Depending on the one or more protocols used to create a pipe, the name can be composed of several protocol-dependent elements.

TCP/IP Routing

The TCP/IP protocol moves information from one computer to another by means of *packets*. Packets are groups of data including the actual information to be transmitted and routing information. The routing information is used to guide the packet from its source computer to its destination computer.

For a packet to reach its destination, the following sequence of events must occur:

1. A client opens a connection to a computer by contacting a *DNS (Domain Name System)* or *WINS (Windows Internet Name Service)* server to resolve a hostname to an IP address.

2. The DNS or WINS server consults one of several possible information sources to convert a text *hostname* into a binary IP address.

3. The DNS or WINS server sends back the binary IP address to the client computer, usually as a value in a *WinSock* function call.

4. The client computer sends its packet, including the binary IP address obtained in Step 3, to the next computer on the network, either the client's gateway or another peer machine depending on network topology.

5. The next computer on the network determines whether that IP address is intended for itself; if so, it keeps the packet and sends an acknowledgment that it did so. Otherwise, the packet is sent to the next appropriate computer, until its destination is found.

The preceding sequence of events is called *routing*. Routing can be accomplished by using a gateway computer or a dedicated router and a set of routing tables. The routing tables are used to actually make the forwarding decisions based on the network topology.

DHCP

DHCP stands for *Dynamic Host Configuration Protocol*. DHCP allows computers that might be on a network only part-time to share dedicated IP addresses. It can also assist network administrators by reducing the overhead of manually resetting hard-allocated IP addresses when a network topology reorganizes. DHCP functions as a specialized DNS (see previous section) that works within a very limited domain, usually a small *LAN (Local Area Network)* or intranet. The DHCP server accepts a hostname just like a DNS server, looks it up in its internal data elements, and returns a binary IP address. What makes DHCP useful is that—unlike standard DNS and WINS servers—it is optimized for easy mapping rearrangements between names and binary addresses.

DNS

The *Domain Name System (DNS)* resolves "human-friendly" computer names (such as **www.coriolis.com**) to TCP/IP addresses. This makes it possible for systems to find each other in LANs and WANs (wide area networks). The use of DNS is much more common in Windows 2000 networks since Microsoft opted for TCP/IP as the default protocol and actually requires the use of DNS for Active Directory implementations. Remember, Active Directory is the new directory service of W2K. This directory service permits scalability, integration, and optimization of the operating system like never before. The DNS is actually a network itself. If one DNS server does not know the answer to a query, it is configured to ask another DNS server in the network.

WINS

The *Windows Internet Name Service (WINS)* also resolves human-friendly computer names to TCP/IP addresses. Unfortunately, WINS is a proprietary Microsoft technology and Microsoft is trying to abandon this methodology in favor of the more standard and recognized DNS technology. Thanks to all of the many networks that still use earlier Microsoft operating systems, WINS will be around for quite some time.

Many other networking topics are necessary for clearly understanding how to tune a Windows 2000 computer using the Registry. You should have a thorough understanding of Windows NT/2000 networking essentials before you attempt to make the adjustments described in the "Immediate Solutions" section. Entire multivolume sets have been written about Windows NT networking essentials, some of which might prove to be useful references when exercising the options in the "Immediate Solutions" section.

6. TCP/IP and the Internet

Immediate Solutions

Forcing SNAP Encoding on TCP/IP Routing Packets

SNAP encoding is a way to reduce the size of packets for transmission over Ethernet. To make sure your TCP/IP network is using this feature, tweak the Registry as follows:

1. Launch Regedt32.

2. Select the Window menu option for **HKEY_LOCAL_MACHINE**.

3. Use the tree control in the left-hand window to navigate to the **SYSTEM\CurrentControlSet\Services\Tcpip\Parameters** subkey. Double-click on the subkey to expand it.

4. Locate the **ArpUseEtherSNAP** value entry. Use the DWORD Editor to change its value to 1 and force the use of SNAP encoding.

TIP: If the ArpUseEtherSNAP value is not present, you can add it with the Edit\Add value menu option.

Setting the Path for TCP/IP Database Files

The TCP/IP system requires a set of special text files to help keep track of the various parts of the system (based on the Unix text-file system for which WinSock was developed). These files include HOSTS, LMHOSTS, and SERVICES. The location of the text files is in the Parameters subkey in the Registry. You can access the Parameters subkey by following these steps:

1. Launch Regedt32.

2. Select the Window menu option for **HKEY_LOCAL_MACHINE**.

3. Use the tree control in the left-hand window to navigate to the **SYSTEM\CurrentControlSet\Services\Tcpip\Parameters** subkey. Double-click on the subkey to expand it.

4. Locate the **DatabasePath** value entry. Set its REG_SZ string value to the directory pathname where you would like these files stored.

Setting the DNS Server Address for a TCP/IP Server

The **DHCPNameServer** parameter specifies the DNS name servers that Windows Sockets is to query to resolve names. The DHCP client service—if enabled—writes this parameter. If the **NameServer** parameter has a valid value, it overrides this parameter:

1. Launch Regedt32.
2. Select the Window menu option for **HKEY_LOCAL_MACHINE**.
3. Use the tree control in the left-hand window to navigate to the **SYSTEM\CurrentControlSet\Services\Tcpip\Parameters** subkey. Double-click on the subkey to expand it.
4. Locate the **DhcpNameServer** value entry. Set its REG_SZ string value to the IP address or addresses of the DNS servers.

6. TCP/IP and the Internet

Enabling Dead Gateway Detection for a TCP/IP Server

If your network has multiple gateways, you can quickly switch automatically from one that goes down to another, functional one using Windows 2000 features. To do this, first enter all the available gateways in the Advanced section of the TCP/IP configuration dialog. Then, tweak the Registry as follows:

1. Launch Regedt32.
2. Select the Window menu option for **HKEY_LOCAL_MACHINE**.
3. Use the tree control in the left-hand window to navigate to the **SYSTEM\CurrentControlSet\Services\Tcpip\Parameters** subkey. Double-click on the subkey to expand it.
4. Locate the **EnableDeadGWDetect** value entry. Use the DWORD Editor to change the EnableDeadGWDetect value to 1 to enable dead gateway detection, and set

EnableDeadGWDetect to 0 to disable it. Enabling dead gateway detection causes a round-robin search among all listed gateways until a gateway is found that accepts packets.

Increasing Buffer Memory for TCP/IP Routing Performance

The **ForwardBufferMemory** parameter determines how much memory IP allocates initially to store packet data in the router packet queue. When this buffer space is filled, the system attempts to allocate more memory. Packet queue data buffers are 256 bytes in length, so the value of this parameter should be a multiple of 256. Multiple buffers are chained together for larger packets. The IP header for a packet is stored separately. This parameter is ignored and no buffers are allocated if the IP routing function is not enabled. The maximum amount of memory that can be allocated for this function is controlled by **MaxForwardBufferMemory**. Setting this value too high, wastes memory. Setting the value to low hurts TCP/IP routing performance. Follow these steps to set this value:

1. Launch Regedt32.

2. Select the Window menu option for **HKEY_LOCAL_MACHINE**.

3. Use the tree control in the left-hand window to navigate to the **SYSTEM\CurrentControlSet\Services\Tcpip\Parameters** subkey. Double-click on the subkey to expand it.

4. Locate the **ForwardBufferMemory** value entry. Use the DWORD Editor to change its value to a multiple of 1,480, rounded up to be evenly divisible by 256. The multiples of 1,480 are the number of 50-packet chunks the buffer can hold at one time.

Changing **GlobalMaxTcpWindowSize**

Windows 2000, unlike all previous versions of NT, supports large windows as described in RFC1323. (*RFC* stands for *Request for Comment.*) This can really help speed up TCP/IP traffic transfers. **GlobalMaxTcpWindowSize** is a DWORD value in the Registry with a range of 0 to 1073741824. If you are using your system with DSL or a cable modem, you should consider a setting of 256960. No matter what setting you use, for the best results GlobalMaxTcpWindowSize should be a multiple of *MSS (Maximum Segment Size)*. MSS is generally MTU - 40, where *MTU (Maximum Transmission Unit)* is the largest packet size that can be transmitted. Follow these instructions to make the change:

1. Launch Regedt32.

2. Select the Window menu option for **HKEY_LOCAL_MACHINE**.

3. Use the tree control in the left-hand window to navigate to the **SYSTEM\CurrentControlSet\Services\Tcpip\Parameters** subkey. Double-click on the subkey to expand it.

4. Set **GlobalMaxTcpWindowSize** to the appropriate value. If this setting is not present, you can add it.

5. In order to be able to set the window size larger than 64KB, you must also make another Registry modification in the same location. Set **Tcp1323Opts** (DWORD) to a value of 3. If this setting is not present, you can add it.

Changing Time to Live (TTL)

Time to live dictates how long a packet can be routed in the network before it is considered impossible to send it to its destination. While it might benefit network performance to set this value lower, be sure not to set it too low. A value that is too low prevents some packets from ever having a chance of reaching their destination. The default value is 128, and in many networks, this is too high. Many reconfigure the time to live to 64. Another widely used value is 32, but this might be too small for a larger network. Here is how you make this Registry change:

1. Launch Regedt32.

2. Select the Window menu option for **HKEY_LOCAL_MACHINE**.

3. Use the tree control in the left-hand window to navigate to the **SYSTEM\CurrentControlSet\Services\Tcpip\Parameters** subkey. Double-click on the subkey to expand it.

4. Set the **DefaultTTL** DWORD value to the appropriate setting.

6. TCP/IP and the Internet

Discovering MTU Automatically

This is another Registry change that can really help out if your system is using one of the fast new Internet connections, such as a cable modem or DSL. When the **EnablePMTUDiscovery** value in the Registry is set to 1, TCP attempts to discover MTU automatically over the path to a remote host. Setting this parameter to 0 causes MTU to default to 576. This typically reduces overall performance over high-speed connections. Here is how you make the change:

1. Launch Regedt32.

2. Select the Window menu option for **HKEY_LOCAL_MACHINE**.

3. Use the tree control in the left-hand window to navigate to the **SYSTEM\CurrentControlSet\Services\Tcpip\Parameters** subkey. Double-click on the subkey to expand it.

4. Ensure the **EnablePMTUDiscovery** value (DWORD) is set to 1 if you want to discover MTU automatically.

Troubleshooting Black Hole Routers

Black hole routers are routers that do not send an "ICMP Destination Unreachable" message when they cannot forward an IP datagram (*ICMP* stands for *Internet Control Message Protocol*). Instead they very rudely just ignore the datagram. Setting the **EnablePMTUBHDetect** value to 1 causes these routers to be detected. Keep in mind that while this is a good thing, setting this value also increases the maximum number of retransmissions for a given segment. To make this change, follow these steps:

1. Launch Regedt32.

2. Select the Window menu option for **HKEY_LOCAL_MACHINE**.

3. Use the tree control in the left-hand window to navigate to the **SYSTEM\CurrentControlSet\Services\Tcpip\Parameters** subkey. Double-click on the subkey to expand it.

4. Ensure the **EnablePMTUBHDetect** value (DWORD) is set to 1 if you want to detect these routers.

Checking Default DHCP Gateway Addresses for a TCP/IP Server

The **DhcpDefaultGateway** parameter specifies the list of default gateways the network is to use to route packets that are not destined for a subnet to which the computer is directly connected and for which a more specific route does not exist. The DHCP client service (if enabled) writes this parameter and a valid **DefaultGateway** parameter value overrides it. Although the parameter is set on a per-interface basis, there is always only one default gateway active for the computer. Additional entries are treated as alternatives if the first one is not available. Follow these steps to set this value:

1. Launch Regedt32.

2. Select the Window menu option for **HKEY_LOCAL_MACHINE**.

3. Use the tree control in the left-hand window to navigate to the **SYSTEM\CurrentControlSet\Services\\[ADAPTERNAME]\Parameters** Tcpip subkey, where *[ADAPTERNAME]* is the network card connecting the host to the DHCP server. Double-click on the subkey to expand it.

4. Locate the **DhcpDefaultGateway** value entry. DhcpDefaultGateway contains multiple strings with all known gateways used for packet routing when the DHCP server cannot locate a client. If a subnet is losing packets, it is very likely that this entry has somehow been removed.

Checking the DHCP Lease Renewal Value for a TCP/IP Client

DHCP uses a *lease system* to assign its IP address pool. Remember, the address pool is the total address space available for TCP/IP clients. An active node must renew its lease after a period of time to keep it; otherwise, the node must apply for a new lease and reset all its internal values—a major performance hit unless the server has been offline. To prevent losing an IP address while a node is active, check the node's renewal timeout entry in the Registry by following these steps:

1. Launch Regedt32.

2. Select the Window menu option for **HKEY_LOCAL_MACHINE**.

6. TCP/IP and the Internet

3. Use the tree control in the left-hand window to navigate to the **SYSTEM\CurrentControlSet\Services\[*ADAPTERNAME*]\Parameters** Tcpip subkey, where **[*ADAPTERNAME*]** is the network card connecting the node to the DHCP server. Double-click on the subkey to expand it.

4. Locate the **T1** value entry. T1 represents the value, in milliseconds after initial assignment, in which the system will try to renew the lease on its IP address from the DHCP server.

Setting the API Protocol for a DHCP TCP/IP Server

DHCP supports three protocols: TCP/IP, Named Pipes, and *Local Procedure Calls (LPCs)*. If you add a client to the network that uses a different routing protocol, but that still needs to use the existing DHCP server, you must edit the Registry as follows:

1. Use Registry Editor to access the Windows Registry.

2. Select the Window menu option for **HKEY_LOCAL_MACHINE**.

3. Use the tree control in the left-hand window to navigate to the **SYSTEM\CurrentControlSet\Services\DhcpServer\Parameters** subkey. Double-click on the subkey to expand it.

4. Locate the **APIProtocolSupport** value entry. Use the DWORD Editor to change APIProtocolSupport to one of the values shown in Table 6.1, depending on the needed protocol support.

Table 6.1   Values used to set DHCP server API protocols.

Value	API Protocols Recognized
1	TCP/IP
2	Named Pipes
4	LPC
5	TCP/IP and LPC
7	TCP/IP, LPC, and Named Pipes

Setting the Path for Backing up DHCP Data for a TCP/IP Server

The DHCP system can be set to log any changes it detects in the dynamic network configuration it serves. These changes are then written to a backup database in case the server crashes before it updates its master information. To change the location of this database file, follow these steps:

1. Launch Regedt32.

2. Select the Window menu option for **HKEY_LOCAL_MACHINE**.

3. Use the tree control in the left-hand window to navigate to the **SYSTEM\CurrentControlSet\Services\DhcpServer\Parameters** subkey. Double-click on the subkey to expand it.

4. Locate the **BackupDatabasePath** value entry. Use the String Editor to change the BackupDatabasePath value to the new directory.

Enabling DHCP Logging for Crash Restoration on a TCP/IP Server

The DHCP system can be set to log any changes it detects in the dynamic network configuration it serves. This log can then be used if the server crashes before it creates its backup database. This facility is on by default, but it imposes a performance penalty. If you need to turn off this feature, here's how to tweak the Registry:

1. Launch Regedt32.

2. Select the Window menu option for **HKEY_LOCAL_MACHINE**.

3. Use the tree control in the left-hand window to navigate to the **SYSTEM\CurrentControlSet\Services\DhcpServer\Parameters** subkey. Double-click on the subkey to expand it.

4. Locate the **DatabaseLoggingFlag** value entry. Use the DWORD Editor to change its value to 0 to disable logging and 1 to enable.

6. TCP/IP and the Internet

Setting the NetBEUI **NameServerPort** Value for a TCP/IP Server

Microsoft WINS listens on Port 0x89, and this is the default configuration used by NetBEUI. If you are using a server that does not use Microsoft WINS, you might need to modify the Registry to change this value:

1. Launch Regedt32.

2. Select the Window menu option for **HKEY_LOCAL_MACHINE**.

3. Use the tree control in the left-hand window to navigate to the **SYSTEM\CurrentControlSet\Services\Netbt\Parameters** subkey. Double-click on the subkey to expand it.

4. Locate the **NameServerPort** value entry. Use the DWORD Editor to change its value to the port number supplied with your custom NetBIOS server implementation.

Setting the Node Type for NetBEUI on a TCP/IP Server

There are several methods NetBEUI can use to resolve names on its system. You can choose which method it uses by altering the Registry as follows:

1. Launch Regedt32.

2. Select the Window menu option for **HKEY_LOCAL_MACHINE**.

3. Use the tree control in the left-hand window to navigate to the **SYSTEM\CurrentControlSet\Services\Netbt\Parameters** subkey. Double-click on the subkey to expand it.

4. Locate the **NodeType** value entry. Use the DWORD Editor to change NodeType to one of the values shown in Table 6.2.

Table 6.2   Values used to set the node types for TCP/IP NetBEUI name resolution.

Value	Node Type	Resolution
1	B-Node	Broadcasts
2	P-Node	WINS Point-to-Point
3	M-Node	Broadcasts, then WINS
4	H-Node	WINS, then Broadcasts

Setting the **Keep-Alive** Value for NetBEUI Sessions

NetBEUI sessions use the *keep-alive value* to determine whether a given node has gone down. You can decrease the interval between checks for a mission-critical network or even disable it entirely. To change the **keep-alive** value via the Registry, follow these steps:

1. Launch Regedt32.

2. Select the Window menu option for **HKEY_LOCAL_MACHINE**.

3. Use the tree control in the left-hand window to navigate to the **SYSTEM\CurrentControlSet\Services\Netbt\Parameters** subkey. Double-click on the subkey to expand it.

4. Locate the **SessionKeepAlive** value entry. Use the DWORD Editor to change its value to a time in milliseconds between keep-alive checks. Setting it to a hex value of **0xFFFFFFFF** disables all keep-alive checking. The default is 3,600,000 (1 hour).

Changing the **Timeout** Value for WINS Name Server Resolution Attempts in NetBEUI on a TCP/IP Server

NetBIOS can make use of a WINS name server, but it might have problems with the requests timing out on a heavily loaded network. Tweak the Registry as follows to overcome this problem:

1. Launch Regedt32.

2. Select the Window menu option for **HKEY_LOCAL_MACHINE**.

3. Use the tree control in the left-hand window to navigate to the **SYSTEM\CurrentControlSet\Services\Netbt\Parameters** subkey. Double-click on the subkey to expand it.

4. Locate the **WinsDownTimeout** value entry. Use the DWORD Editor to change WinsDownTimeout to more than 15,000 (15 seconds) to prevent inap2propriate timeouts.

6. TCP/IP and the Internet

Changing the IP Address Name Server Resolution Order for a TCP/IP Server

Four Registry entries control the order in which the four main categories of name service servers are tried to obtain an IP address from a name string. Here is how to set the Registry to determine the IP address name server resolution order:

1. Launch Regedt32.

2. Select the Window menu option for **HKEY_LOCAL_MACHINE**.

3. Use the tree control in the left-hand window to navigate to the **SYSTEM\CurrentControlSet\Control\Services\Tcpip\ServiceProvider** subkey. Double-click on the subkey to expand it.

4. To change the priority of local name servers, set the **LocalPriority** DWORD value to a new value. Setting a lower value gives a local name server a higher priority for resolution; setting a higher value reduces a local name server's priority.

5. To change the priority of hostname servers, set the **HostsPriority** DWORD value to a new value. Setting a lower value gives a host name server a higher priority for resolution; setting a higher value reduces the host name server's priority.

6. To change the priority of DNS name servers, set the **DnsPriority** DWORD value to a new value. Setting a lower value gives a DNS name server a higher priority for resolution; setting a higher value reduces the DNS name server's priority.

7. To change the priority of NetBT name servers, set the **NetbtPriority** DWORD value to a new value. Setting a lower value gives a NetBT name server a higher priority for resolution; setting a higher value reduces a NetBT name server's priority.

WARNING! Do not set this NetBT name server entry for a remote network to a value less than 1,000, or the network will be swamped with name server requests.

Changing the Excluded Name Service Providers for a TCP/IP Server

Most TCP/IP networks don't use all the possible name servers for their name resolution needs. If a new type of name server becomes available and an old one leaves the system, you will need to perform the following steps to alter the Registry to reflect the change:

1. Launch Regedt32.

2. Select the Window menu option for **HKEY_LOCAL_MACHINE**.

3. Use the tree control in the left-hand window to navigate to the **SYSTEM\CurrentControlSet\Control\Service Providers\ Order** subkey. Double-click the subkey to expand it.

4. Locate the **ExcludedProviders** value entry. Use the Multi-String Editor to Enter and/or delete any of the entries in Table 6.3 (use the number only, not the name). The numbers should be separated by spaces.

WARNING! Remember, entries in the ExcludedProviders value are excluded from name server resolution actions.

Table 6.3   The service numbers used to exclude name server types for TCP/IP sockets.

Service Number	Service
1	NS_SAP
2	NS_NDS
10	NS_TCPIP_LOCAL
11	NS_TCPIP_HOSTS
12	NS_DNS
13	NS_NETBT
14	NS_WINS
20	NS_NBP
30	NS_MS
31	NS_STDA
32	NS_CAIRO
40	NS_X500
41	NS_NIS

6. TCP/IP and the Internet

Changing the Sockets Data Threshold for Flow Control Activation on a TCP/IP Server

The Sockets system has two values it uses to allow straight-through data flow without imposing a flow control algorithm. You can change these values to meet unique local network conditions with the following Registry tweaks:

1. Launch Regedt32.

2. Select the Window menu option for **HKEY_LOCAL_MACHINE**.

3. Use the tree control in the left-hand window to navigate to the **SYSTEM\CurrentControlSet\Afd\Parameters** subkey. Double-click on the subkey to expand it.

4. Locate the **DefaultReceiveWindow** and **DefaultSendWindow** value entries. Use the DWORD Editor to change both their values to above 8,192 to allow more data flow control, or below 8,192 to allow less data flow control due to local line noise.

Changing the Sockets Buffer Allocation to Speed Performance on a TCP/IP Server

Windows Sockets allocates memory buffers for internal use. You can change the Registry to increase the amount of memory allocated and speed WinSock performance as follows:

1. Launch Regedt32.

2. Select the Window menu option for **HKEY_LOCAL_MACHINE**.

3. Use the tree control in the left-hand window to navigate to the **SYSTEM\CurrentControlSet\Afd\Parameters** subkey. Double-click on the subkey to expand it and display its values in the right-hand window.

4. Locate the **InitialSmallBufferCount**, **InitialMedium-BufferCount**, and **InitialLargeBufferCount** value entries, depending on your machine's memory footprint. Use the DWORD Editor to change its value to from 0 through 10 for large buffers, 0 through 30 for medium buffers, and 0 through 50 for small buffers.

TIP: A small memory machine is less than 20MB. A medium memory machine is from 20MB to 125MB. A large memory machine is above 125MB.

Creating a WINS Proxy Agent

You can create a WINS proxy agent out of a Windows 2000 system with a simple Registry modification. A WINS proxy agent listens for broadcasts of non-WINS configured clients and contacts a WINS server on behalf of the client. This can be a very useful feature to add to your network and eliminate unnecessary broadcasts. Follow these instructions to make the change:

1. Launch Regedt32.

2. Select the Window menu option for **HKEY_LOCAL_MACHINE**.

3. Use the tree control in the left-hand window to navigate to the **SYSTEM\CurrentControlSet\Services\Netbt\Parameters** subkey. Double-click on the subkey to expand it and display its values in the right-hand window.

4. If the **EnableProxy** value does not exist (type DWORD), add it to the subkey. A value of 1 causes the system to act as a WINS proxy agent.

Setting the Number of Threads Used by WINS on a TCP/IP Server

The WINS service spawns a number of *worker threads* to help it handle its tasks. Setting the maximum number of such threads is much like an art form heavily dependent on local conditions and resources. The maximum number of threads can be set, only by tweaking the Registry as follows:

1. Launch Regedt32.

2. Select the Window menu option for **HKEY_LOCAL_MACHINE**.

3. Use the tree control in the left-hand window to navigate to the **SYSTEM\CurrentControlSet\Wins\Parameters** subkey. Double-click the subkey to expand it and display its values in the right-hand window.

4. Locate the **NoOfWrkThds** value entry. Use the DWORD Editor to change NoOfWrkThds to a value from 1 through 40.

NOTE: *The default NoOfWrkThds value is the number of processors on the server computer.*

Changing the Process Priority for WINS on a TCP/IP Server

WINS is a process just like any other. Changing its default process priority can sometimes have a salutary effect on network performance. Here's how to alter the WINS default process priority:

1. Launch Regedt32.

2. Select the Window menu option for **HKEY_LOCAL_MACHINE**.

3. Use the tree control in the left-hand window to navigate to the **SYSTEM\CurrentControlSet\Wins\Parameters** subkey. Double-click on the subkey to expand it and display its values in the right-hand window.

4. Locate the **PriorityClassHigh** value entry. Use the DWORD Editor to change the value to 1, to make sure the WINS process does not get preempted by other, less-important processes.

Removing and Reinstalling TCP/IP or Its Services

Let's say you have removed TCP/IP or one of its services from your Windows 2000 system. At a later time, you attempt to reinstall the TCP/IP service. You are troubled when you receive the following error message: "The registry subkey already exists." As you might guess, to correct this problem, you should ensure that all of appropriate Registry subkeys are removed from the Registry.

If you removed TCP/IP and its related service components, you must also remove the following Registry subkeys:

- **HKEY_LOCAL_MACHINE\SYSTEM\CurrentControlSet\ Services\NetBT**

- **HKEY_LOCAL_MACHINE\SYSTEM\CurrentControlSet\ Services\Tcpip**

- **HKEY_LOCAL_MACHINE\SYSTEM\CurrentControlSet\ Services\TcpipCU**

- **HKEY_LOCAL_MACHINE\SYSTEM\CurrentControlSet\ Services\Dhcp**

- **HKEY_LOCAL_MACHINE\SYSTEM\CurrentControlSet\ Services\LmHosts**

If you removed the SNMP (Simple Network Management Protocol) service components, you must also remove the following Registry subkeys:

- **HKEY_LOCAL_MACHINE\SOFTWARE\Microsoft\ RFC1156Agent**

- **HKEY_LOCAL_MACHINE\SOFTWARE\Microsoft\Snmp**

- **HKEY_LOCAL_MACHINE\SYSTEM\CurrentControlSet\ Services\Snmp**

If you removed the TCP/IP Printing service components, you must also remove the following Registry subkeys:

- **HKEY_LOCAL_MACHINE\SOFTWARE\Microsoft\Lpdsvc**

- **HKEY_LOCAL_MACHINE\SOFTWARE\Microsoft\ TcpPrint**

- **HKEY_LOCAL_MACHINE\SYSTEM\CurrentControlSet\ Services\ LpdsvcSimple TCP/IP Services**

If you removed the Simple TCP/IP Services components, you must also remove the following Registry subkeys:

- **HKEY_LOCAL_MACHINE\SOFTWARE\Microsoft\ SimpTcp**

- **HKEY_LOCAL_MACHINE\SYSTEM\CurrentControlSet\ Services\SimpTcp**

If you removed the DHCP service components, you must also remove the following Registry subkeys:

- **HKEY_LOCAL_MACHINE\SOFTWARE\Microsoft\ DhcpMibAgent**

- **HKEY_LOCAL_MACHINE\SOFTWARE\Microsoft\ DhcpServer**

- **HKEY_LOCAL_MACHINE\SYSTEM\CurrentControlSet\ Services\DhcpServer**

If you removed the WINS service components, you must also remove the following Registry subkeys:

- **HKEY_LOCAL_MACHINE\SOFTWARE\Microsoft\Wins**

- **HKEY_LOCAL_MACHINE\SOFTWARE\Microsoft\ WinsMibAgent**

- **HKEY_LOCAL_MACHINE\SYSTEM\CurrentControlSet\ Services\Wins**

6. TCP/IP and the Internet

If you have removed the DNS service components, you must also remove the following Registry subkeys:

- **HKEY_LOCAL_MACHINE\SOFTWARE\Microsoft\Dns**

- **HKEY_LOCAL_MACHINE\SOFTWARE\Microsoft\ DnsMibAgent**

- **HKEY_LOCAL_MACHINE\SYSTEM\CurrentControlSet\ Services\Dns**

Chapter 7

Hardware and Systems

(continued)

In Brief

Windows 2000 is composed of a number of systems that work together to provide the functionality of the OS. These systems include the *Hardware Abstraction Layer (HAL)*, the *Win16 on Win32 (WOW)* system, and various system services. Services are applications that the administrator adds to the OS and that are not strictly part of Windows 2000 (although several services are provided with it). Each of these systems has a number of important Registry entries that are the focus of this chapter.

Hardware Abstraction Layer (HAL)

Each operating system, regardless of its design, must have some way to connect the actual hardware it uses with the software that needs it. For an application to write to the screen, save a file to disk, or get input from the mouse, it must be able to interact with the computer hardware. In Windows 2000, the collection of device drivers that provides this capability is called the Hardware Abstraction Layer (HAL).

The HAL provides a number of additional services beyond simply initializing device drivers, including the following:

- Creating and maintaining settings for both serial and bus mice, including startup parameters, event queue size, and resolution
- Maintaining information on non-SCSI (Small Computer Systems Interface) hard drives, including size, number of heads, tracks, and cylinders
- Creating identifier strings for serial and parallel ports, as well as maintaining their IRQ (Interrupt Request) and access settings
- Handling SCSI debugging capabilities, including disconnects, synchronous transfers, multiple connections, and tagged connections
- Providing installation and system file paths for the OS itself, as well as its version and installation date

7. Hardware and Systems

WOW (Win16 on Win32)

Windows NT has always provided the capability to run MS-DOS and Windows 3.x applications, and Windows 2000 continues this capability. The technologies that run MS-DOS and Windows 3.x applications are called WOW (Win16 on Win32) and the *DOS VM (virtual machine)*.

MS-DOS

To handle MS-DOS applications, Windows 2000 loads a copy of Autoexec.bat and Config.sys on startup. Problems with these files won't affect the successful starting of Windows 2000, but they can impact MS-DOS applications running under Windows 2000.

WOW and INI Files

To handle Windows 3.x applications, Windows 2000 reads a copy of each INI file it finds during initial installation or the installation of a Windows 3.x application into the Windows 2000 Registry. There are ways to force a refresh of these Registry values from their INI file sources each time a new user logs in. WOW runs as a system service. This can have drawbacks, because it might require "killing" a phantom Windows 3.x application to allow Windows 2000 to shut down. A Registry tweak can fix this behavior.

Other System Services

Services are applications that execute as part of the Windows 2000 operating system. Microsoft provides some, such as Internet Information Server (IIS) and SQL Server. (SQL stands for Structured Query Language.) Other services are acquired by administrators for specific needs and written by third-party developers. Regardless of their source, services are all administered in the same way. They are placed in a startup queue and executed in the order listed in the queue. If problems are encountered, error message boxes pop up or errors are logged to the event logs. After the OS starts all services, it runs any applications in its startup group. Finally, when the OS is ready to shut down, a specific timeout value is provided to allow administrators to abort the shutdown if necessary.

Immediate Solutions

Controlling Response to Service Startup Failure

When Windows 2000 encounters an error attempting to start a service, it can respond automatically in several different ways. As you might guess, you can control exactly how Windows 2000 responds using a very important Registry value. To control this setting, do the following:

1. Launch Regedt32.

2. Select the Window menu option for **HKEY_LOCAL_MACHINE**.

3. Navigate to the **SYSTEM\CurrentControlSet\Services\ *[DriverName]*** subkey in the Registry and double-click on the subkey to expand it.

4. Add the **ErrorControl** DWORD value if it is not present.

5. Configure the **ErrorControl** value using the potential values listed in Table 7.1 below:

Table 7.1   Possible **ErrorControl** values and their effect.

Value	Meaning	Effect
0	Ignore	Startup continues normally—no error reported
1	Normal	Startup continues normally—warning reported
2	Severe	**LastKnownGood** control set used to boot the system; if it is already in use the system ignores the error
3	Critical	**LastKnownGood** control set used to boot the system; if it is already in use, the startup process stops and an error is reported

7. Hardware and Systems

Controlling the Start Order of Windows 2000 Services

Although Windows 2000 itself normally determines the order in which its services are started, an administrator can modify this by changing the Registry as follows:

1. Launch Regedt32.

2. Select the Window menu option for **HKEY_LOCAL_MACHINE**.

3. Use the tree control in the left-hand window to navigate to the **SYSTEM\CurrentControlSet\Control\ServiceGroupOrder** subkey. Double-click on the subkey to expand it and display its value in the right pane.

4. Locate the **List** value entry. Use the Multi-String Editor to examine the order in which specific service groups are loaded. Changing the order of the groups in the list changes the order in which drivers are loaded and services are started. For a typical Windows 2000 Server configuration, the default start order is as follows:

    -   System Reserved
    -   Boot Bus Extender
    -   System Bus Extender
    -   SCSI Miniport
    -   Port
    -   Primary Disk
    -   SCSI Class
    -   SCSI CDROM Class
    -   Filter
    -   Boot File System
    -   Base
    -   Pointer Port
    -   Keyboard Port
    -   Pointer Class
    -   Keyboard Class
    -   Video Init
    -   Video

- Video Save
- File System
- Event Log
- Streams Drivers
- NDIS (Network Device Interface Specification) Wrapper
- PNP_TDI
- NDIS
- TDI (Transport Driver Interface)
- NetBIOSGroup
- PlugPlay (PNP)
- SpoolerGroup
- NetDDEGroup
- Parallel Arbitrator
- Extended Base
- RemoteValidation
- PCI (Peripheral Component Interconnect) Configuration
- MS Transactions

Determining Whether a Mouse Uses a Serial or Bus Connection

An administrator can determine whether a mouse is using serial or bus connections by examining the Registry as follows:

1. Launch Regedt32.
2. Select the Window menu option for **HKEY_LOCAL_MACHINE**.
3. Use the tree control in the left-hand window to navigate to **HARDWARE\Description\System\MultifunctionAdapter\\[X]\\Pointer-Controller\\[X]\\PointerPeripheral\\0**, where *[X]* is the number containing subkeys. (This is usually 0, but it can be higher; for example, it is 25 on our sample server.) Double-click on the subkey to expand it and display its values in the right-hand window.
4. Locate the **Identifier** REG_SZ value entry. If it is empty, the mouse is a bus mouse; if the Identifier entry contains a value, the mouse is a serial mouse.

Obtaining Non-SCSI Hard Disk Information

Although some Windows 2000 computers use SCSI hard drives, some now use large IDE (Integrated Development Environment) or EIDE (Enhanced Integrated Development Environment) devices. These devices might need their BIOS (Basic Input/Output System) settings changed. To obtain detailed hardware information about the devices, follow these steps:

1. Launch Regedt32.

2. Select the Window menu option for **HKEY_LOCAL_MACHINE**.

3. Use the tree control in the left-hand window to navigate to the **HARDWARE\Devicemap\AtDisk\Controller[#]\Disk[#]** subkey, where **Controller[#]** and **Disk[#]** are the assigned numbers for the IDE controller card and the specific hard disk on the card. Double-click on the subkey to select it and display its values in the right-hand window. The values shown in Table 7.2 reflect the hardware capabilities of specific hard drives.

Table 7.2   Non-SCSI hard drive values.

Value	Information
Firmware revision	REG_SZ value with the revision number of the BIOS code
Identifier	REG_SZ value with the manufacturer's disk identifier
Number of cylinders	REG_DWORD with the number of cylinders on the drive
Number of heads	REG_DWORD with the number of heads on the drive
Sectors per track	REG_DWORD in hexadecimal with the number of sectors on one track of the drive, typically 512

Checking the Video Driver Used by Windows 2000

Many times in Windows 2000, you can have a problem loading a particular video driver on a system. Typically, this is because some program has replaced some files on your Windows 2000 system. Symptoms of this problem are usually:

- The system continues to default to VGA (Video Graphics Array) mode or only boots in VGA mode from the boot menu.

- Your video driver is not loading properly.

In order to specify exactly which video drivers Windows 2000 is to load for the video system, follow these steps:

1. Launch Regedt32.

2. Select the Window menu option for **HKEY_LOCAL_MACHINE**.

3. Navigate to the following location:
   HKEY_LOCAL_MACHINE\hardware\DeviceMap\Video.

4. Double-click the subkey to expand it. The **Device\Video0** value points to a Registry key that has the location of the video driver Windows 2000 is configured to load.

5. To see the value setting, double-click on the value **InstalledDisplayDrivers**. This is the driver Windows 2000 has loaded.

6. If it is set to another Registry location, a third-party driver may be being loaded when Windows 2000 starts. For example, if PcAnywhere is installed, Device0 does not have the name of a file, but instead displays another Registry location that gives the video driver name.

Setting a Parallel Port Identification String

You can change the identifier for a given *parallel port* from its system-assigned value. This could assist you in advanced troubleshooting techniques when you need to easily identify the port. To do this, tweak the Registry as follows (remember, you have to perform the following steps each time the machine is rebooted because the entries are volatile):

1. Launch Regedt32.

2. Select the Window menu option for **HKEY_LOCAL_MACHINE**.

3. Use the tree control in the left-hand window to navigate to the **HARDWARE\Devicemap\ParallelPorts** subkey. Double-click on the subkey to expand it and display its values in the right-hand window.

4. Locate the **Parallel[X]:** value entry, where *[X]* is the port number you want to change. Use the String Editor to change the *Parallel[X]:* value to the new identifier (the default is *LPT[X]:*).

7. Hardware and Systems

Setting a Parallel Port Access Level

You can change the access level for a given parallel port from its system-assigned value. To do this, tweak the Registry as follows:

1. Launch Regedt32.

2. Select the Window menu option for **HKEY_LOCAL_MACHINE**.

3. Use the tree control in the left-hand window to navigate to the **SYSTEM\CurrentControlSet\Services\Parallel\Parameters** subkey. Double-click on the subkey to expand it and display its values in the right-hand window.

4. If a subkey named **Parallel[X]:** (where *[X]* is the port number of interest) does not exist, enter one. Add a value entry for this subkey (if one is not present) named **DisablePort**, of type REG_DWORD. To enable access after a port is initialized, set DisablePort to 0. To disable access after a port is initialized, set DisablePort to 1.

Setting a Parallel Port IRQ Value

You can change the IRQ for a given parallel port from its system-assigned value. To do this, tweak the Registry as follows:

1. Launch Regedt32.

2. Select the Window menu option for **HKEY_LOCAL_MACHINE**.

3. Use the tree control in the left-hand window to navigate to the **SYSTEM\CurrentControlSet\Services\Parallel\Parameters** subkey. Double-click on the subkey to expand it and display its values in the right-hand window.

4. If a subkey named **Parallel[X]:** (where *[X]* is the port number of interest) does not exist, enter one. Add a value entry for this subkey (if one does not exist) named **Interrupt**, of type REG_DWORD. Set **Interrupt** to the IRQ entry desired.

Setting a Serial Port Identification String

You can change the identifier for a given *serial port* from its system-assigned value. To do this, tweak the Registry as follows:

1. Launch Regedt32.

2. Select the Window menu option for **HKEY_LOCAL_MACHINE**.

3. Use the tree control in the left-hand window to navigate to the HARDWARE\Devicemap\Serial\ports subkey. Double-click on the subkey to expand it and display its values in the right-hand window.

4. Locate the **Serial[X]:** value entry, where **[X]** is the port number you want to change. Use the String Editor to change the **Serial[X]:** value to the new identifier (the default is **COM[X]:**).

Setting a Serial Port Access Level

You can change the access level for a given serial port from its system-assigned value. To do this, tweak the Registry as follows:

1. Launch Regedt32.

2. Select the Window menu option for **HKEY_LOCAL_MACHINE**.

3. Use the tree control in the left-hand window to navigate to the **SYSTEM\CurrentControlSet\Services\Serial\Parameters** subkey. Double-click on the subkey to expand it and display its values in the right-hand window.

4. If a subkey named **Serial[X]:** (where **[X]** is the port number of interest) does not exist, enter one. Then, add a value entry for this subkey (if one does not exist) named **DisablePort**, of type REG_DWORD. To enable access after a port is initialized, set DisablePort to 0. To disable access after a port is initialized, set DisablePort to 1.

Setting a Serial Port IRQ Value

You can change the IRQ for a given serial port from its system-assigned value. To do this, tweak the Registry as follows:

1. Launch Regedt32.

2. Select the Window menu option for **HKEY_LOCAL_MACHINE**.

3. Use the tree control in the left-hand window to navigate to the **SYSTEM\CurrentControlSet\Services\Serial\Parameters**

7. Hardware and Systems

sub-key. Double-click on the subkey to expand it and display its values in the right-hand window.

4. If a new subkey named **Serial[X]:** (where *[X]* is the port number of interest) does not exist, enter one. Then, add a value entry for this subkey (if one does not exist) named **Interrupt**, of type REG_DWORD. Set **Interrupt** to the IRQ entry desired.

Enabling a Serial Port FIFO Queue

Serial ports can support FIFO (first in, first out) queuing—rather than LIFO (last in, first out), the default in Windows 2000. To turn FIFO queuing on for a given serial port, tweak the Registry as follows:

1. Launch Regedt32.

2. Select the Window menu option for **HKEY_LOCAL_MACHINE**.

3. Use the tree control in the left-hand window to navigate to the **SYSTEM\CurrentControlSet\Services\Serial\Parameters** subkey. Double-click on the subkey to expand it and display its values in the right-hand window.

4. If a subkey named **Serial[X]:** (where *[X]* is the port number of interest) does not exist, enter one. Then, add a value entry for the subkey (if one does not exist) named **ForceFifoEnable**, of type REG_DWORD. Set ForceFifoEnable to 1 to force FIFO queuing for a given port.

WARNING! FIFO queuing is not always reliable. If data loss is experienced on the port, disable the FIFO queuing feature.

Setting the Size of a Bus Mouse Event Queue

Windows 2000 sets a default amount of memory to hold (or queue) mouse events. In some cases (such as high-resolution monitors with a fine-grained mouse resolution), events might be lost, resulting in error messages in the error log file. To prevent these errors, change the Registry as follows:

1. Launch Regedt32.

2. Select the Window menu option for **HKEY_LOCAL_MACHINE**.

3. Use the tree control in the left-hand window to navigate to the **SYSTEM\CurrentControlSet\Services\Busmouse\Parameters** subkey. Double-click on the subkey to expand it and display its values in the right-hand window.

4. Locate the **MouseDataQueueSize** value entry. Use the DWORD Editor to change the MouseDataQueueSize value (in hexadecimal) to a new, higher value. This number is the total mouse events that can be queued, and finding the proper number might be a matter of trial and error.

Setting the Resolution of a Bus Mouse

Windows 2000 sets a default *sample rate* to check the mouse, per second. For a high-resolution screen, the default sample rate might be too low and mouse movements might be missed. To prevent mouse movement errors for a computer using a bus mouse, change the Registry as follows:

1. Launch Regedt32.

2. Select the Window menu option for **HKEY_LOCAL_MACHINE**.

3. Use the tree control in the left-hand window to navigate to the **SYSTEM\CurrentControlSet\Services\Busmouse\Parameters** subkey. Double-click on the subkey to expand it and display its values in the right-hand window.

4. Locate the **SampleRate** value entry. Use the DWORD Editor to change the SampleRate value (in hexadecimal) to a new, higher value. This number is the number of times per second the mouse is checked, and finding the proper number might be a matter of trial and error.

Setting the Size of a Serial Mouse Event Queue

Windows 2000 sets a default amount of memory to hold (or queue) for mouse events. In some cases (such as high-resolution monitors with a fine-grained mouse resolution), events might be lost, resulting in error messages in the error log file. To prevent mouse event errors for a computer using a serial mouse, change the Registry as follows:

1. Launch Regedt32.

2. Select the Window menu option for **HKEY_LOCAL_MACHINE**.

3. Use the tree control in the left-hand window to navigate to the **SYSTEM\CurrentControlSet\Services\Sermouse\Parameters** subkey. Double-click on the subkey to expand it and display its values in the right-hand window.

4. Locate the **MouseDataQueueSize** value entry. Use the DWORD Editor to change the MouseDataQueueSize value (in hexadecimal) to a new, higher value. This number is the total mouse events that can be queued, and finding the proper number might be a matter of trial and error.

Setting the SCSI Debugging Level

SCSI is a hard disk controller system often used with Windows 2000 hard drives. Windows 2000 offers a set of debugging features for situations in which such drives experience problems. Turning on SCSI debugging requires tweaking the Registry as follows:

1. Launch Regedt32.

2. Select the Window menu option for **HKEY_LOCAL_MACHINE**.

3. Use the tree control in the left-hand window to navigate to the **SYSTEM\CurrentControlSet\Services*[DEVICENAME]*\Parameters** subkey, where *[DEVICENAME]* is the name of the SCSI adapter in question. Double-click on the subkey to expand it and display its values in the right-hand window.

4. Locate the **ScsiDebug** value entry. Use the DWORD Editor to change the ScsiDebug value to a setting from 0 (the default, with no error messages) through 3 (full error messages). This setting enables the system to display error messages based on the other SCSI debugging settings.

Activating SCSI Driver Entry Debug Breakpoints

You can also set a SCSI breakpoint in the debugging process. This requires tweaking the Registry as follows:

1. Launch Regedt32.

2. Select the Window menu option for **HKEY_LOCAL_MACHINE**.

3. Use the tree control in the left-hand window to navigate to the **SYSTEM\CurrentControlSet\Services\\[*DEVICENAME*]\Parameters** subkey, where *[DEVICENAME]* is the name of the SCSI adapter in question. Double-click on the subkey to expand it and display its values in the right-hand window.

4. Locate the **BreakPointOnEntry** value entry. Use the DWORD Editor to change the BreakPointOnEntry value to 1 to enable breaking inside **SpParseDevice** by the system debugger. Set the BreakPointOnEntry value to 0 to disable this feature.

Disabling SCSI Disconnects During Debugging

To facilitate debugging SCSI devices, you can force them to execute their disk I/O serially rather than in parallel; this requires tweaking the Registry as follows:

1. Launch Regedt32.

2. Select the Window menu option for **HKEY_LOCAL_MACHINE**.

3. Use the tree control in the left-hand window to navigate to the **SYSTEM\CurrentControlSet\Services\\[*DEVICENAME*]\Parameters** subkey, where *[DEVICENAME]* is the name of the SCSI adapter in question. Double-click on the subkey to expand it and display its values in the right-hand window.

4. Locate the **DisableDisconnects** value entry. Use the DWORD Editor to change DisableDisconnects to 1 to disable SCSI bus disconnects; this causes all requests to be executed sequentially. Set the DisableDisconnects value back to 0 to disable this feature.

7. Hardware and Systems

Disabling SCSI Multiple Requests during Debugging

To prevent more than one request at a time being sent to a SCSI device, edit the Registry as follows:

1. Launch Regedt32.

2. Select the Window menu option for **HKEY_LOCAL_MACHINE**.

3. Use the tree control in the left-hand window to navigate to the **SYSTEM\CurrentControlSet\Services\\[DEVICENAME]\Parameters** subkey, where *[DEVICENAME]* is the name of the SCSI adapter in question. Double-click on the subkey to expand it and display its values in the right-hand window.

4. Locate the **DisableMultipleRequests** value entry. Use the DWORD Editor to change DisableMultipleRequests to 1 to disable sending more than one request at a time for a given SCSI device (the default setting). Set the DisableMultipleRequests value to 0 to disable this feature during debugging.

Disabling SCSI Synchronous Transfers during Debugging

The SCSI bus normally transfers data synchronously, which can interfere with debugging. To disable that behavior, change the Registry as follows:

1. Launch Regedt32.

2. Select the Window menu option for **HKEY_LOCAL_MACHINE**.

3. Use the tree control in the left-hand window to navigate to the **SYSTEM\CurrentControlSet\Services\\[DEVICENAME]\Parameters** subkey, where *[DEVICENAME]* is the name of the SCSI adapter in question. Double-click on the subkey to expand it and display its values in the right-hand window.

4. Locate the **DisableSynchronousTransfers** value entry. Use the DWORD Editor to set DisableSynchronousTransfers to 1 to disable synchronous transfers over the SCSI bus (the default). Set the DisableSynchronousTransfers value to 0 to disable this feature during debugging.

Enabling SCSI-II Tagged Command Queueing During Debugging

You can turn off tagging of SCSI commands in the queue with the following Registry edit:

1. Launch Regedt32.

2. Select the Window menu option for **HKEY_LOCAL_MACHINE**.

3. Use the tree control in the left-hand window to navigate to the **SYSTEM\CurrentControlSet\Services*DEVICENAME*\Parameters** subkey, where *[DEVICENAME]* is the name of the SCSI adapter in question. Double-click on the subkey to expand it and display its values in the right-hand window.

4. Locate the **DisableTaggedQueueing** value entry. Use the DWORD Editor to change the DisableTaggedQueueing value to 1 to disable tagging of commands in the queue (the default). Set the DisableTaggedQueueing value to 0 to enable command tagging during debugging, provided the SCSI adapter supports SCSI-II features.

Getting the Installed Windows 2000 Version Information

Some applications check the Registry for information about the installed version of Windows 2000. If an application is refusing to run for no obvious reason, you can check whether the Registry entry for the Windows 2000 version number is valid. You also might need to check this important version information for other reasons as well. Check version information by following these steps:

1. Launch Regedt32.

2. Select the Window menu option for **HKEY_LOCAL_MACHINE**.

3. Use the tree control in the left-hand window to navigate to the **SOFTWARE\Microsoft\WindowsNT\CurrentVersion** subkey. Double-click on the subkey to expand it and display its values in the right-hand window.

4. Locate the **CurrentBuild**, **CurrentType**, and **CurrentVersion** value entries. Use the DWORD Editor to check these values against the documentation values for the troublesome application; if they are wrong, change them to the proper values.

7. Hardware and Systems

Related solution:	Found on page:
Determining the Version of a Printer's Driver	200

Getting the Installed Windows 2000 System Path Information

Some applications check the Registry for information about the installation path for Windows 2000. You can easily check the Registry for this information that is reported to applications by following these steps:

1. Launch Regedt32.

2. Select the Window menu option for **HKEY_LOCAL_MACHINE**.

3. Use the tree control in the left-hand window to navigate to the **SOFTWARE\Microsoft\WindowsNT\CurrentVersion** subkey. Double-click on the subkey to expand it and display its values in the right-hand window.

4. Locate the **SystemRoot** and **PathName** value entries. Use the String Editor to check the value of each entry against the documentation value for the troublesome application; if it is wrong, change it to the proper value.

WARNING! The SystemRoot and PathName values are not found in the Environment Registry entries.

Related solution:	Found on page:
Setting the Path to a Printer's Data File	199

Getting the Installed Windows 2000 Setup Path Information

Windows 2000 "remembers" the path from which you installed W2K and will default to this path when you subsequently attempt to add or remove Windows 2000 components. To update this path information in the Registry, do the following:

1. Launch Regedt32.

2. Select the Window menu option for **HKEY_LOCAL_MACHINE**.

7. Hardware and Systems

3. Use the tree control in the left-hand window to navigate to the **SOFTWARE\Microsoft\WindowsNT\CurrentVersion** subkey. Double-click on the subkey to expand it and display its values in the right-hand window.

4. Locate the **SourcePath** value entry. Use the String Editor to check and/or modify the SourcePath value.

Changing the Programs to Run at Windows 2000 Startup

An administrator can add or remove programs to be run at Windows 2000 startup. These programs are executed before services are executed on the system. To add or remove startup programs, follow these steps:

1. Launch Regedt32.

2. Select the Window menu option for **HKEY_LOCAL_MACHINE**.

3. Use the tree control in the left-hand window to navigate to the **SYSTEM\CurrentControlSet\Control\Session Manager** subkey. Double-click on the subkey to expand it and display its values in the right-hand window.

4. Locate the **BootExecute** value entry. Use the Multi-String Editor to add the command line for an application to be run or to remove the command line to stop an application from running.

Removing a Bad Startup Job Entry

One of the more excruciating errors for Windows 2000 is the one that indicates that a job required by the system cannot be run. This normally happens when an application changes a Registry entry so its one-time installation utility will run on reboot after it installs and then it either crashes or is badly written and does not remove the Registry entry. Here's how to deal with this problem:

1. Launch Regedt32.

2. Select the Window menu option for **HKEY_CURRENT_USER**.

3. Use the tree control in the left-hand window to navigate to the **SOFTWARE\Microsoft\WindowsNT\CurrentVersion\Windows** subkey. Double-click on the subkey to expand it and display its values in the right-hand window.

4. Locate the **Load** and **Run** value entries if they exist. Use the String Editor to make sure the Load and Run values are not set to nonexistent or invalid entries.

Suppressing NonSystem Startup Error Popup Windows

A possible annoyance of Windows 2000 is its tendency to pop up error dialog boxes during bootup and wait for a user response. This can cause havoc if there isn't an administrator on duty to get rid of them. Fortunately, you can suppress noncritical error popups; here's how to adjust the Registry:

1. Launch Regedt32.

2. Select the Window menu option for **HKEY_CURRENT_USER**.

3. Use the tree control in the left-hand window to navigate to the **SOFTWARE\Microsoft\WindowsNT\CurrentVersion\Windows** subkey. Double-click on the subkey to expand it and display its values in the right-hand window.

4. Add an **ErrorMode** REG_DWORD entry if one does not exist. Then, set the ErrorMode value to 1 to suppress nonsystem error message popups.

Suppressing All Startup Error Popup Windows

In order to eliminate all startup error messages, follow these steps:

1. Launch Regedt32.

2. Select the Window menu option for **HKEY_CURRENT_USER**.

3. Use the tree control in the left-hand window to navigate to the **SOFTWARE\Microsoft\WindowsNT\CurrentVersion\Windows** subkey. Double-click on the subkey to expand it and display its values in the right-hand window.

7. Hardware and Systems

4. Add an **ErrorMode** REG_DWORD entry if one does not exist. Set the ErrorMode value to 2 to suppress all error message popups.

Setting the Windows 2000 Shutdown Time

You might experience problems in Windows 2000 with the system shutting down before a given service can save its data, resulting in inconsistent behavior. Or perhaps you are having a related, but opposite problem. You have few services running or those that you are using are shutting down quickly and efficiently. Because Windows 2000 gives each service up to 20 seconds to shut down by default before the system steps in and kills it, this time might add up on your system and cause shutdowns to be painfully slow. Here is how you can control these situations:

1. Launch Regedt32.

2. Select the Window menu option for **HKEY_LOCAL_MACHINE**.

3. Use the tree control in the left-hand window to navigate to the **SYSTEM\CurrentControlSet\Control** subkey. Double-click on the subkey to expand it and display its values in the right-hand window.

4. Locate the **WaitToKillServiceTimeout** value entry. Use the DWORD Editor to change the WaitToKillServiceTimeout value to a higher or lower value in milliseconds (the default is 20,000, or 20 seconds) so that the offending service has enough time to shut down.

Handling Spurious Hotfix Error Messages

A real hair-puller can happen when you attempt to use a Hotfix for Windows 2000 and it blithely informs you that it is already installed, when it isn't. Instead of checking for specific ID values under a Registry key, a Hotfix might simply check for the existence of the key itself! Here's how to handle this tooth-grinder:

1. Launch Regedt32.

2. Select the Window menu option for **HKEY_LOCAL_MACHINE**.

3. Use the tree control in the left-hand window to navigate to the **SOFTWARE\Microsoft\WindowsNT\CurrentVersion** subkey. Double-click on the subkey to expand it and display its values in the right-hand window.

4. Delete the Hotfix subkey. This allows the errant Hotfix to install itself and re-create the key.

Handling Startup A: Drive Error Messages

Windows 2000 machines that run DOS applications or Windows 3.x programs can sometimes display weird error messages stating that there is no disk in the floppy drive during bootup. This is due to a problem with the Autoexec.bat file's **PATH** statements. Here's how to change the Registry to kill this little nuisance:

1. Launch Regedt32.

2. Select the Window menu option for **HKEY_LOCAL_MACHINE**.

3. Use the tree control in the left-hand window to navigate to the **SYSTEM\SETUP** subkey. Double-click on the subkey to expand it and display its values in the right-hand window.

4. Locate the **WinntPath** value entry. Use the String Editor to scan the WinntPath string for invalid path entries to the A: drive (or, for that matter, a CD-ROM drive). If the invalid path entry does not exist, then the problem is happening on another level.

Setting WOW Device Not Ready Timeouts

Some networked devices might experience problems with Device Not Ready messages using Windows 3.x or MS-DOS applications. To deal with this problem, you can change the WOW Registry entry as follows:

1. Launch Regedt32.

2. Select the Window menu option for **HKEY_LOCAL_MACHINE**.

3. Use the tree control in the left-hand window to navigate to the **SOFTWARE\Microsoft\WindowsNT\CurrentVersion** subkey. Double-click on the subkey to expand it and display its values in the right-hand window.

4. Locate the **DeviceNotSelectedTimeout** value entry. Use the DWORD Editor to change the DeviceNotSelectedTimeout value to a higher number (the default is 15) in seconds to account for the slow peripheral. If this value does not exist—you can create it.

Preventing WOW Windows 2000 Shutdown Aborts

A really delightful error can happen with WOW preventing Windows 2000 from shutting down until a nonexistent Windows 16 process is killed. This little gem can be eliminated as follows:

1. Launch Regedt32.

2. Select the Window menu option for **HKEY_USERS**.

3. Use the tree control in the left-hand window to navigate to the **DEFAULT\Control Panel\Desktop** subkey. Double-click on the subkey to expand it and display its values in the right-hand window.

4. If a value entry named **AutoEndTasks**, of type REG_SZ, is not already present, create one. Set the AutoEndTasks value to 1 to force WOW to shut down without the annoying "fake task" error.

7. Hardware and Systems

Chapter 8

General Networking

In Brief

Windows 2000's networking capabilities are dependent on a group of systems usually referred to as General Networking Services:

- Workstation
- Browser
- UPS (uninterruptible power supply)
- Alerter
- Server
- Directory Replication
- Messenger

Workstation

Windows 2000's *Workstation* service is the General Networking Service that controls how clients connect to the operating system. The Workstation service controls how many aspects of Windows 2000's behavior, including the following:

- Named Pipes behavior
- Datagram propagation and error recovery
- Performance enhancements, such as lock-read and opportunistic locking (oplock)
- Raw I/O performance enhancements
- Closed file caching
- Open file caching
- Session and user limits

Browser

In order for computers on a Windows 2000 network to share their resources, the *Browser service* must be enabled and functioning correctly. The Browser service's main function is to keep each computer on the network aware of changes made to shared resources on other computers. This service has been almost completely replaced thanks to the inclusion of Active Directory; however, this is still a necessary service in most configurations.

UPS

The *UPS service* handles uninterruptible power supplies, which are special batteries that come into operation the instant normal power is interrupted. This prevents data loss and allows the administrator or a script to safely shut down the machine. This includes permitting the service to "gracefully" shut down the system once the power outage is detected.

Alerter

The *Alerter service* is an intermediary between other services and the administrator. The administrator configures an administrative alert for a particular behavior or state of some other service. When the occurrence of that behavior triggers the alert, the system logs an event via the Messenger service to inform the administrator.

Server

The *Server service* deals with Remote Procedure Calls (RPCs), file sharing, printer sharing, and Named Pipe sharing. Its main job is to coordinate between those lower-level services and system resources they might need.

Immediate Solutions

Setting Users to Receive Administrator Alerts

Even the best networks experience problems. Windows 2000 uses the Alerter service to send system alert messages when needed via the Messenger service. To add or remove users to receive alerts, tweak the Registry as follows:

1. Launch Regedt32.

2. Select the Window menu option for **HKEY_LOCAL_MACHINE**.

3. Use the tree control in the left-hand window to navigate to the **SYSTEM\CurrentControlSet\Services\Alerter\Parameters** subkey. Double-click on the subkey to expand it and display its values in the right-hand window.

4. Locate the **AlertNames** value entry. Use the Multi-String Editor to add and remove usernames that appear on the list of users who should receive system alert messages.

Enhancing Named-Pipes Performance with Character Buffering

The Workstation service is the primary system used to connect clients to Windows servers. One Registry tweak regarding the Workstation service is to enable buffering of character-mode Named Pipes (thus speeding up Named Pipe performance). This should be done only if the application's documentation indicates that buffering its Named Pipes will not cause a problem. To enable Named Pipe buffering, follow these steps:

1. Launch Regedt32.

2. Select the Window menu option for **HKEY_LOCAL_MACHINE**.

3. Use the tree control in the left-hand window to navigate to the **SYSTEM\CurrentControlSet\Services\LanmanWorkstation\ Parameters** subkey. Double-click on the subkey to expand it and display its values in the right-hand window.

4. Locate the **BufNamedPipes** value entry. Set its REG_DWORD value to 1 to enable buffering and 0 to disable it. If this value is not present, you can create it.

TIP: *If you need to make sure that all pipe write operations are flushed immediately to the server, disable the BufNamedPipes entry. Be sure to consult the documentation for all applications using Named Pipes to ensure this does not cause them to break.*

Enhancing Lock-Read Performance by Increasing Buffer Size

The **LockQuota** value specifies the maximum amount of data that is read for each file if the **UseLockReadUnlock** parameter is enabled. You should consider increasing this value if your Named Pipes application performs a significant number of lock-read style operations. (This means performing lock operations and immediately reading the contents of the locked data.)

1. Launch Regedt32.

2. Select the Window menu option for **HKEY_LOCAL_MACHINE**.

3. Use the tree control in the left-hand window to navigate to the **SYSTEM\CurrentControlSet\Services\LanmanWorkstation** Parameters subkey. Double-click on the subkey to expand it and display its values in the right-hand window.

4. Locate the **LockQuota** value entry. Set its REG_DWORD value to a higher value as needed. Obviously, you should only increase this value up to the optimal performance level for your system. If this value does not exist, you can create it.

WARNING! *You might run out of paged pool memory if you set the LockQuota value too high (in the megabytes range).*

8. General Networking

Enhancing Named-Pipes Performance by Increasing the Pipe Throughput

You can tweak the Registry with respect to the Workstation service to increase the number of characters sent through a Named Pipe before the write is buffered.

WARNING! *Administrators must consult their developers or documentation to ensure that such a change will not break existing applications.*

To make this adjustment, follow these steps:

1. Launch Regedt32.
2. Select the Window menu option for **HKEY_LOCAL_MACHINE**.
3. Use the tree control in the left-hand window to navigate to the **SYSTEM\CurrentControlSet\Services\LanmanWorkstation\Parameters** subkey. Double-click on the subkey to expand it and display its values in the right-hand window.
4. Locate the **MaxCollectionCount** value entry. Set its REG_DWORD value to a higher value in bytes to decrease the amount of Named-Pipe write buffering and to a lower value to increase buffering. If this value is not present, you can create it.

TIP: The minimum value for MaxCollectionCount is 0; the maximum is 65,535.

Enhancing Data Integrity by Enabling Lock-Read-Write-Unlock Optimizations

UseLockReadUnlock indicates whether the redirector uses the lock-read and write-and-unlock performance enhancements. When this value is enabled, it generally provides a significant performance benefit. However, database applications that lock a range and do not allow data within that range to be read suffer performance degradation unless this parameter is disabled. To do this follow these steps:

1. Launch Regedt32.
2. Select the Window menu option for **HKEY_LOCAL_MACHINE**.

3. Use the tree control in the left-hand window to navigate to the **SYSTEM\CurrentControlSet\Services\LanmanWorkstation\Parameters** subkey. Double-click on the subkey to expand it and display its values in the right-hand window.

4. Locate the **UseLockReadUnlock** value entry. Set its REG_DWORD value to 1 to enable this performance enhancement and 0 to disable it. If this value is not present, you can create it.

Enhancing Performance by Enabling Opportunistic Locking

You can also tweak the Registry with regard to the Workstation service to enable opportunistic locking (oplock). This should be done only if all applications that use the Workstation service indicate they can use it; otherwise, it might disable an application. To enabling opportunistic locking, follow these steps:

1. Launch Regedt32.

2. Select the Window menu option for **HKEY_LOCAL_MACHINE**.

3. Use the tree control in the left-hand window to navigate to the **SYSTEM\CurrentControlSet\Services\LanmanWorkstation\Parameters** subkey. Double-click on the subkey to expand it and display its values in the right-hand window.

4. Locate the **UseOpportunisticLocking** value entry. Set its REG_DWORD value to 1 to enable this performance enhancement and 0 to disable it. If this value does not exist, you can create it.

Enhancing Performance by Enabling Raw-Read Optimization

You can also tweak the Registry with respect to the Workstation service to enable raw-read operations (highly useful on a high-bandwidth LAN). To enable raw-read operations, follow these steps:

1. Launch Regedt32.

2. Select the Window menu option for **HKEY_LOCAL_MACHINE**.

8. General Networking

3. Use the tree control in the left-hand window to navigate to the **SYSTEM\CurrentControlSet\Services\LanmanWorkstation\ Parameters** subkey. Double-click on the subkey to expand it and display its values in the right-hand window.

4. Locate the **UseRawRead** value entry. Set its REG_DWORD value to 1 to enable this performance enhancement and 0 to disable it. If the value does not exist you can create it.

TIP: *You should also enable raw writes when you enable UseRawRead. Use the same technique described in this Immediate Solution, but set the UseRawWrite entry to 1.*

Enhancing Performance by Enabling Raw-Write-With-Data Optimizations

Another Registry tweak with respect to the Workstation service is enabling raw-write-with-data optimization. This allows the redirector to send 4KB of data with each write-raw operation, which provides a significant performance enhancement on a local area network. To make this change, follow these steps:

1. Launch Regedt32.

2. Select the Window menu option for **HKEY_LOCAL_MACHINE**.

3. Use the tree control in the left-hand window to navigate to the **SYSTEM\CurrentControlSet\Services\LanmanWorkstation\ Parameters** subkey. Double-click on the subkey to select it and display its values in the right-hand window.

4. Locate the **UseWriteRawData** value entry. Set its REG_DWORD value to 1 to enable this performance enhancement and 0 to disable it. If the value does not exist you can create it.

WARNING! *You must have already enabled raw writes to use the UseWriteRawData enhancement. This is described in another Immediate Solution in this Chapter.*

Enhancing Named-Pipes Performance by Reducing Availability Delay

To reduce the delay between the availability of a Named Pipe and its creation, you can tweak the Registry regarding the Workstation service. However, under some circumstances, this can cause data loss or corruption, so it should be used carefully. To enable this performance enhancement, follow these steps:

1. Launch Regedt32.

2. Select the Window menu option for **HKEY_LOCAL_MACHINE**.

3. Use the tree control in the left-hand window to navigate to the **SYSTEM\CurrentControlSet\Services\LanmanWorkstation\ Parameters** subkey. Double-click on the subkey to expand it and display its values in the right-hand window.

4. Locate the **CharWait** value entry. Set its REG_DWORD value to a lower value to decrease the wait time and a higher value to increase the wait time.

TIP: You can increase the CharWait value when your pipe server application becomes very busy, then, decrease CharWait when the load eases off.

Preventing Named-Pipe Server Swamping

The **PipeMaximum** Registry tweak regarding the Workstation service controls the maximum time at which the redirector "backs off" on failing nonblocking pipe reads. To make this enhancement, follow these steps:

1. Launch Regedt32.

2. Select the Window menu option for **HKEY_LOCAL_MACHINE**.

3. Use the tree control in the left-hand window to navigate to the **SYSTEM\CurrentControlSet\Services\LanmanWorkstation\ Parameters** subkey. Double-click on the subkey to expand it and display its values in the right-hand window.

4. Locate the **PipeMaximum** value entry. Set its REG_DWORD value to a lower value to prevent server swamping from nonblocking Named Pipes. This value can be added if it does not already exist. The default value for this setting is 500 (milliseconds).

TIP: *The backoff statistics can be used to fine-tune the PipeMaximum entry.*

Enhancing Performance by Increasing the Cache Time for Closed Files

Increasing the amount of time a closed file is left in the file cache allows quicker reopening of the same file. However, you must use this Registry tweak with care, because too long a cache time can cause data loss under certain conditions. To change the length of time a closed file remains in the file cache, follow these steps:

1. Launch Regedt32.

2. Select the Window menu option for **HKEY_LOCAL_MACHINE**.

3. Use the tree control in the left-hand window to navigate to the **SYSTEM\CurrentControlSet\Services\LanmanWorkstation\ Parameters** subkey. Double-click on the subkey to expand it and display its values in the right-hand window.

4. Locate the **CacheFileTimeout** value entry. Set its REG_DWORD value to a higher value to keep closed files in the cache longer; the value is indicated in seconds. If the value does not exist, you can create it. The default is 10 seconds.

TIP: *A situation in which to increase the CacheFileTimeout value is when your users will be performing programming builds over the network.*

Enhancing Performance by Increasing the Total Cached Closed Files

Thanks to another Workstation Registry entry, you can specify the maximum number of files that should be left open on a share after the application has closed a particular file. This parameter exists because the default configuration of LAN Manager servers allow a total of only 60 open files from remote clients and 50 from each client workstation. Because the Windows 2000 redirector can keep files open in the cache after an application has closed the file, this means that the redirector can overload a misconfigured LAN Manager server. To correct this problem, either reduce this value or increase the values for the LAN Manager server's **maxSessopens** and **maxOpens** parameters:

1. Launch Regedt32.

2. Select the Window menu option for **HKEY_LOCAL_MACHINE**.

3. Use the tree control in the left-hand window to navigate to the **SYSTEM\CurrentControlSet\Services\LanmanWorkstation\Parameters** subkey. Double-click on the subkey to expand it and display its values in the right-hand window.

4. Locate the **DormantFileLimit** value entry. Set its REG_DWORD value to a higher value to keep more closed files in the cache. If the value does not exist you can create it. The default is 45.

TIP: *It is a good idea to increase the DormantFileLimit value if your users are going to be performing programming builds over the network.*

Fixing Lost Mailslot Message Errors

You can tweak the Registry with respect to the Workstation service to increase the available buffer for mailslots and prevent lost messages. It is important to realize that this might well result in a performance hit and should be done only if there is a clear issue involving lost messages that cannot be resolved in other ways. To do this, follow these steps:

1. Launch Regedt32.

2. Select the Window menu option for **HKEY_LOCAL_MACHINE**.

3. Use the tree control in the left-hand window to navigate to the **SYSTEM\CurrentControlSet\Services\LanmanWorkstation\Parameters** subkey. Double-click on the subkey to expand it and display its values in the right-hand window.

4. Locate the **MailslotBuffers** value entry. Set its REG_DWORD value to a higher value to prevent losing mailslot messages. You can add this value if you need to.

8. General Networking

TIP: *The MailslotBuffers default value is 5.*

Disabling Windows 2000 File Caching

This Workstation-related tweak indicates whether the redirector uses the cache manager to cache the contents of files. You should disable this parameter only to guarantee that all data is flushed to the server immediately after it is written by the application.

To do this, follow these steps:

1. Launch Regedt32.

2. Select the Window menu option for **HKEY_LOCAL_MACHINE**.

3. Use the tree control in the left-hand window to navigate to the **SYSTEM\CurrentControlSet\Services\LanmanWorkstation\ Parameters** subkey. Double-click the subkey to expand it and display its values in the right-hand window.

4. Locate the **UtilizeNtCaching** value entry. Set its REG_DWORD value to 1 to enable caching and 0 to disable it. You can add this value if you need to.

TIP: *Disable the UtilizeNtCaching entry only if you are having a problem with flushing data to the server.*

Enabling Raw I/O Performance Enhancements

The Server service is the primary system used to connect clients to shared resources. One Registry tweak with respect to the Server service is increasing the number of raw I/O work items available. This can lead to improved performance for large file transfers. To do this, follow these steps:

1. Launch Regedt32.

2. Select the Window menu option for **HKEY_LOCAL_MACHINE**.

3. Use the tree control in the left-hand window to navigate to the **SYSTEM\CurrentControlSet\Services\LanmanServer\Parameters** subkey. Double-click on the subkey to expand it and display its values in the right-hand window.

4. Locate the **RawWorkItems** value entry. Set its REG_DWORD value to a higher number to increase the throughput of raw I/O. The value can be from 1 through 512. This value can be added if it does not exist.

WARNING! *Setting the RawWorkItems value too high can degrade performance because of increased memory usage.*

Controlling the Maximum I/O Clients

Another valuable Registry tweak with regard to the Server service is setting a hard value for the number of simultaneous users allowed to log on to a given Windows 2000 Server.

To do this, follow these steps:

1. Launch Regedt32.
2. Select the Window menu option for **HKEY_LOCAL_MACHINE**.
3. Use the tree control in the left-hand window to navigate to the **SYSTEM\CurrentControlSet\Services\LanmanServer\Parameters** subkey. Double-click on the subkey to expand it and display its values in the right-hand window.
4. Locate the **Users** value entry. Set its REG_DWORD value to the number of users that is your desired threshold for the given Windows 2000 Server computer. This value can be added if it does not exist.

TIP: *To allow infinite users, set the Users value to* **0xFFFFFFFF.**

Controlling Oplock Break Behavior

The Server service is the principal system for connecting clients to shared resources. One Registry tweak relating to the Server service is setting how the system behaves when an opportunistic locking encounters an error condition.

WARNING! *Tweak this entry only if any applications on the system will not be impacted negatively. Administrators should consult documentation or developer support to confirm this.*

8. General Networking

To configure how a system behaves when it encounters an oplock, follow these steps:

1. Launch Regedt32.

2. Select the Window menu option for **HKEY_LOCAL_MACHINE**.

3. Use the tree control in the left-hand window to navigate to the **SYSTEM\CurrentControlSet\Services\LanmanServer\Parameters** subkey. Double-click on the subkey to expand it and display its values in the right-hand window.

4. Locate the **EnableOplockForceClose** value entry. Set its REG_DWORD value to 1 to force closing of the file so the new client can access it; setting its REG_DWORD value to 0 keeps the file locked and prevents a new client from accessing it. This value can be added if it does not exist.

WARNING! Setting the EnableOplockForceClose value to 1 can, in some cases, result in the loss of cached data for the oplocked file.

Enabling Oplocking Performance Enhancements

Another Registry tweak regarding the Server service is to enable opportunistic locking on the server (as opposed to workstations). This can be a dramatic performance enhancement, yet should be used cautiously in WAN (wide area network) environments. To do this, follow these steps:

1. Launch Regedt32.

2. Select the Window menu option for **HKEY_LOCAL_MACHINE**.

3. Use the tree control in the left-hand window to navigate to the **SYSTEM\CurrentControlSet\Services\LanmanServer\Parameters** subkey. Double-click the subkey to expand it and display its values in the right-hand window.

4. Locate the **EnableOplocks** value entry. Set its REG_DWORD value to 1 to enable opportunistic locking and 0 to disable it. Add this value if it does not exist.

Fixing Raw I/O Link Delay Timeouts

Another Registry tweak regarding the Server service is to prevent raw I/O timeouts over slow connections. To do this, follow these steps:

1. Launch Regedt32.

2. Select the Window menu option for **HKEY_LOCAL_MACHINE**.

3. Use the tree control in the left-hand window to navigate to the **SYSTEM\CurrentControlSet\Services\LanmanServer\Parameters** subkey. Double-click on the subkey to expand it and display its values in the right-hand window.

4. Locate the **MaxLinkDelay** value entry. Add it if it does not exist. Set its REG_DWORD value to a higher value to prevent slow connections from having raw I/O disabled for them. The default is 60.

TIP: *The MaxLinkDelay value has a range from 0 through 100,000 seconds.*

Enabling Nonpaged Memory Quotas

Windows 2000 automatically sets a default limit for allocable nonpaged pool memory. This default value is approximately 80 percent of installed memory. If the system reaches this limit as a result of network activity, problems can result. To change this limit, modify the Registry as follows:

1. Launch Regedt32.

2. Select the Window menu option for **HKEY_LOCAL_MACHINE**.

3. Use the tree control in the left-hand window to navigate to the **SYSTEM\CurrentControlSet\Services\LanmanServer** subkey. Double-click on the subkey to expand it and display its values in the right-hand window.

4. Locate the **MaxNonpagedMemoryUsage** value entry. Set its REG_DWORD value to a value from 1 to the number of megabytes for the nonpaged memory pool (per user). If this value does not exist, you can create it.

TIP: *To disable memory quotas, enter **0xFFFFFFFF** (infinite) in the MaxNonpagedMemoryUsage value entry.*

Related solution:	Found on page:
Changing the Paged Memory Pool Size	83

Enabling Paged Memory Quotas

Another Registry tweak regarding the Server service is to enable a paged memory quota for individual users. To do this, follow these steps:

1. Launch Regedt32.

2. Select the Window menu option for **HKEY_LOCAL_MACHINE**.

3. Use the tree control in the left-hand window to navigate to the **SYSTEM\CurrentControlSet\Services\LanmanServer\Parameters** subkey. Double-click on the subkey to expand it and display its values in the right-hand window.

4. Locate the **MaxPagedMemoryUsage** value entry. Set its REG_DWORD value to a value from 1 to the number of megabytes for the paged memory pool (per user). If this value does not exist, you can create it.

TIP: *To disable memory quotas, enter **0xFFFFFFFF** (infinite) in the MaxPagedMemoryUsage value entry.*

Related solution:	Found on page:
Changing the Nonpaged Memory Pool Size	83

Changing Server Thread Priority

Another Registry tweak regarding the Server service is to change server threads' priority values. To do this, follow these steps:

1. Launch Regedt32.

2. Select the Window menu option for **HKEY_LOCAL_MACHINE**.

3. Use the tree control in the left-hand window to navigate to the **SYSTEM\CurrentControlSet\Services\LanmanServer\Parameters** subkey. Double-click on the subkey to expand it and display its values in the right-hand window.

Table 8.1   Setting server thread priority levels.

Value	Priority Level
0	Normal thread priority
1	Foreground thread priority
2	Background thread priority
15	Realtime thread priority (not recommended!)

4. Locate the **ThreadPriority** value entry. If this value does not exist, you can create it. Set its REG_DWORD value to one of the values listed in Table 8.1.

WARNING! Changes to this Registry entry may impact system performance substantially.

Handling UPS Startup Failures

If the UPS (uninterruptible power supply) service for Windows 2000 refuses to run, chances are that the UPS service is relying on bad Registry data. To check the UPS Registry values, follow these steps:

1. Launch Regedt32.

2. Select the Window menu option for **HKEY_LOCAL_MACHINE**.

3. Use the tree control in the left-hand window to navigate to the **SYSTEM\CurrentControlSet\Services\UPS** subkey. Double-click on the subkey to expand it and display its values in the right-hand window.

4. Check each value entry against what you have verified actually exists. Change any offending values to the proper values.

5. You can also check the Config subkey for specific UPS configuration settings that may be giving you a problem.

Detecting Slow Network Connections

By default, Windows 2000 systems attempt to download profile information from a Domain Controller. This is fine, except in an environment where there is not suitable bandwidth to transfer the profile to the local system in a reasonable time. Then this feature turns into a nightmare, as logon tends to take forever. Fortunately, you can tweak the Registry and have Windows 2000 detect whether a link is slow or not. Set the threshold for a link to be considered slow. To do this, follow these steps:

1. Launch Regedt32.

2. Select the Window menu option for **HKEY_LOCAL_MACHINE**.

3. Use the tree control in the left-hand window to navigate to the SYSTEM\Software\Microsoft\Windows **NT\CurrentVersion\Winlogon** subkey. Double-click on the subkey to expand it and display its values in the right-hand window.

4. Add a REG_DWORD value named **SlowLinkDetectEnabled** and set the value to 1.

5. Add a REG_DWORD value named **SlowLinkTimeOut** and set it to the number of milliseconds Windows 2000 should wait for a PING response from the Domain Controller.

Chapter 9

Networking Protocols and Interoperability

(continued)

In Brief

Networking in Windows 2000 has two main features: the protocols (including TCP/IP, NetBIOS, IPX/SPX, and DLC) and interoperability, the ability to run various networking APIs (AppleTalk and Novell) on top of the protocols. Windows 2000 networking is strongly tied to Registry entries, which allows a knowledgeable administrator to both fine-tune and optimize networking capabilities with a few well chosen changes.

Networking Protocols

Protocols are the plumbing of LANs and WANs. They control how information is actually moved along the physical wires of a network. Windows 2000 supports a number of protocols, including the following:

- *NDIS (Network Device Interface Specification)*—NDIS is the generic specification for NICs (network interface cards) from Microsoft.

- *TCP/IP (Transmission Control Protocol/Internet Protocol)*— TCP/IP is the Internet's protocol, which is supported for Windows 2000.

- *IPX/SPX (Internet Packet Exchange/Sequenced Packet Exchange)*—Novell's networking protocol, supported in Windows 2000.

- *NetBIOS (Network Basic Input-Output System)*—IBM's network protocol, supported in Windows 2000.

- *NWNBLink (NetWare to NT Binary Link)*—NWNBLink is the API (Application Programming Interface) for the IPX/SPX Novell protocols.

- *NBF (NetBIOS Frame Format)*—An enhanced version of NetBIOS Extended User Interface (NetBEUI), the API for IBM Networks on Windows 2000.

- *DLC (Data Link Control)*—DLC is a special protocol used for communication with mainframe computers and some HP printers.

Networking Interoperability

Windows 2000 can run several networking APIs, including Novell, Windows LAN Manager, AppleTalk, and the Macintosh File (MacFile) system, on top of its protocols. The Registry holds a number of key settings that influence how interoperability functions. The most important ones have "Immediate Solutions" for them in this chapter.

Novell

Novell's NetWare product has been around since the earliest days of MS-DOS. Microsoft provides compatibility for NetWare starting with Windows 3, up to and including Windows 2000. Windows 2000 also includes performance-enhancing extensions that can be used when two Windows 2000 machines are running a NetWare LAN.

Windows LAN Manager

Windows 2000 provides a basic LAN system—Windows LAN Manager—which handles basic networking tasks, such as sharing files and providing access to shared resources such as modems and printers. It also enforces security restrictions and can interact with several different file systems, including FAT (File Allocation Table) and NTFS (New Technology File System).

AppleTalk and the Macintosh File System

Windows 2000 supports the AppleTalk LAN system for Macintosh and the Macintosh file system. This support allows Windows 2000 machines to share files with Macintosh nodes, and vice versa.

9. Networking Protocols and Interoperability

Immediate Solutions

Setting Novell Frame Window Values

NetWare NetBIOS Link (NWNBLink) contains Microsoft enhancements to NetBIOS. The NWNBLink component can format NetBIOS-level requests and pass them to the NWLink component for transmission on the network. You can tweak the Registry to set the number of frames that NWNBLink sends before sending an acknowledgment (which can greatly speed up performance). To perform this tweak, follow these steps:

1. Launch Regedt32.

2. Select the Window menu option for **HKEY_LOCAL_MACHINE**.

3. Use the tree control in the left-hand window to navigate to the **SYSTEM\CurrentControlSet\Services\NWlnkNb\Parameters** subkey. Double-click on the subkey to expand it and display its values in the right-hand window.

4. Locate the **AckWindow** value entry. If this value is not present you can add it. Use the DWORD Editor to change the AckWindow value to 0 to turn off acknowledgments (this assumes all computers are on a fast LAN) or set AckWindow to a value for the number of frames to accept prior to sending an acknowledgment. The default value for this setting is 2.

Enabling Novell Piggybacking

You can tweak the Registry to enable *acknowledgment piggybacking*, which allows attaching an acknowledgment request onto the end of a NetBIOS message rather than sending the request as a separate message. To enable acknowledgment piggybacking, alter the Registry as follows:

1. Launch Regedt32.

2. Select the Window menu option for **HKEY_LOCAL_MACHINE**.

3. Use the tree control in the left-hand window to navigate to the **SYSTEM\CurrentControlSet\Services\NWlnkNb\Parameters**

9. Networking Protocols and Interoperability

subkey. Double-click on the subkey to expand it and display its values in the right-hand window.

4. Locate the **EnablePiggyBackAck** value entry. If the value does not exist, you can create it. Use the DWORD Editor to change the EnablePiggyBackAck value to 1 to enable acknowledgment piggybacking or 0 to disable the feature.

Enabling NWNBLink Extensions

You can also tweak the Registry to enable the various NWNBLink extensions that are necessary for using the frame window settings enhancements and acknowledgment piggybacking discussed in the two previous Immediate Solutions. To enable the NWNBLink extensions, follow these steps:

1. Launch Regedt32.

2. Select the Window menu option for **HKEY_LOCAL_MACHINE**.

3. Use the tree control in the left-hand window to navigate to the **SYSTEM\CurrentControlSet\Services\NWlnkNb\Parameters** subkey. Double-click on the subkey to expand it and display its values in the right-hand window.

4. Locate the **Extensions** value entry. Use the DWORD Editor to change the Extensions value to 1 to enable NWNBLink Extensions or 0 to disable the feature. If the value does not exist, you can add it.

Enabling RIP Routing for IPX

Windows 2000 can function as a RIP (Routing Information Protocol) router for the IPX (Internet Packet Exchange) protocol. Follow these steps to find out where to find this and other important IPX protocol settings controlled in the Registry:

1. Launch Regedt32.

2. Select the Window menu option for **HKEY_LOCAL_MACHINE**.

3. Use the tree control in the left-hand window to navigate to the **SYSTEM\CurrentControlSet\Services\NwLnkIpx\Parameters** subkey. Double-click on the subkey to expand it and display its values in the right-hand window.

4. Locate the **EnableWANRouter** value entry. Use the DWORD Editor to change the EnableWANRouter value to a value of 1 to enable RIP routing on the system. The default value is 0.

Creating a Dedicated Router

If you are going to use your Windows 2000 system solely as a RIP router for the IPX protocol, you can configure the system as a dedicated router. This can dramatically improve the performance of the system. You should configure the system as a dedicated router only if it will not be running other network services, of course. In order to make this change, follow these steps:

1. Launch Regedt32.

2. Select the Window menu option for **HKEY_LOCAL_MACHINE**.

3. Use the tree control in the left-hand window to navigate to the **SYSTEM\CurrentControlSet\Services\NwLnkIpx\Parameters** subkey. Double-click on the subkey to select it and display its values in the right-hand window.

4. Locate the **DedicatedRouter** value entry. Use the DWORD Editor to change the DedicatedRouter value to 1.

Controlling RIP Routing with **RipCount**

If you are using your Windows 2000 system for RIP routing over IPX, you can control routing details to effect improve performance. One such setting is **RipCount**. This controls how many times the system will send a message attempting to find a route on the network before giving up. The default is 5. The possible range is 1 to 65535. To do this perform the following:

1. Launch Regedt32.

2. Select the Window menu option for **HKEY_LOCAL_MACHINE**.

3. Use the tree control in the left-hand window to navigate to the **SYSTEM\CurrentControlSet\Services\NwLnkIpx\Parameters** subkey. Double-click on the subkey to select it and display its values in the right-hand window.

4. Locate the **RipCount** value entry. Use the DWORD Editor to change the value.

9. Networking Protocols and Interoperability

Related solution:	Found on page:
Controlling RIP Routing with **RipTimeout**	178

Controlling RIP Routing with **RipTimeout**

Another important RIP router setting that you can adjust in the Registry is **RipTimeout**. **RipTimeout** is related to **RipCount** and controls the amount of time between packets being sent out while the system is trying to find a route to a destination. The DWORD value defaults to 1 and is measured in half-seconds. The value can be set between 1 and 65535. To control this setting, follow these steps:

1. Launch Regedt32.

2. Select the Window menu option for **HKEY_LOCAL_MACHINE**.

3. Use the tree control in the left-hand window to navigate to the **SYSTEM\CurrentControlSet\Services\NwLnkIpx\Parameters** subkey. Double-click on the subkey to expand it and display its values in the right-hand window.

4. Locate the **RipTimeout** value entry. Use the DWORD Editor to change the RipTimeout value.

Controlling the RIP Cache

The IPX protocol maintains a RIP cache in order to help locate systems on remote networks. You can control how long the IPX protocol keeps an entry in this cache before it is removed. Your options are between 1 minute and 65535 minutes. The default for this value is 15 minutes. To control the value, follow these steps:

1. Launch Regedt32.

2. Select the Window menu option for **HKEY_LOCAL_MACHINE**.

3. Use the tree control in the left-hand window to navigate to the **SYSTEM\CurrentControlSet\Services\NwLnkSpx\Parameters** subkey. Double-click on the subkey to expand it and display its values in the right-hand window.

4. Locate the **RipUsageTime** value entry. Use the DWORD Editor to change the value.

Setting AppleTalk Logon Messages

Windows 2000 supports both the Macintosh file system and the AppleTalk networking protocol. You can use the Registry to set the logon message for clients who connect to the network using the MacFile system. To do this, follow these steps:

1. Launch Regedt32.

2. Select the Window menu option for **HKEY_LOCAL_MACHINE**.

3. Use the tree control in the left-hand window to navigate to the **SYSTEM\CurrentControlSet\Services\MacFile\Parameters** subkey. Double-click on expand to select it and display its values in the right-hand window.

4. Locate the **LoginMsg** value entry. Use the String Editor to change the LoginMsg value to a value appropriate for the current system. If the value does not exist, you can create it. It is a REG_SZ type and can be 1 to 198 characters.

Enabling MacFile Server Performance Enhancements

You can use the Registry to enable several performance enhancements regarding the Macintosh file system, such as maximum simultaneous sessions, paged cache memory pool limits, and nonpaged cache memory pool limits. To activate these enhancements, follow these steps:

1. Launch Regedt32.

2. Select the Window menu option for **HKEY_LOCAL_MACHINE**.

3. Use the tree control in the left-hand window to navigate to the **SYSTEM\CurrentControlSet\Services\MacFile\Parameters** subkey. Double-click on the subkey to expand it and display its values in the right-hand window. Table 9.1 shows three value entries that have a powerful ability to enhance MacFile performance. If any of these values do not exist—they can be added.

9. Networking Protocols and Interoperability

Table 9.1   Setting MacFile performance enhancements.

Registry Key	Value Type	Format	Description
MaxSessions	REG_DWORD	0xFFFFFFFF unlimited	Maximum or number users at one time
PagedMemLimit	REG_DWORD	0x3E8 to 0x3E800 in KB	Maximum Paged File Cache Memory
NonPagedMemLimit	REG_DWORD	0xFF to 0x3E80 in KB	Maximum Nonpaged File Cache Memory

Allowing MacFile Server Guest Logons

You can use the Registry to enable MacFile and AppleTalk guest logons by following these steps:

1. Launch Regedt32.

2. Select the Window menu option for **HKEY_LOCAL_MACHINE**.

3. Use the tree control in the left-hand window to navigate to the **SYSTEM\CurrentControlSet\Services\MacFile\Parameters** subkey. Double-click on the subkey to expand it and display its values in the right-hand window.

4. Locate the **ServerOptions** value entry. Use the DWORD Editor to change the ServerOptions value by adding 1 to the current setting to enable guest logons or subtracting 1 to disable them.

Allowing MacFile Server Clear-Text Passwords

You can use the Registry to allow clear-text passwords. To allow clear-text passwords, follow these steps:

1. Launch Regedt32.

2. Select the Window menu option for **HKEY_LOCAL_MACHINE**.

3. Use the tree control in the left-hand window to navigate to the **SYSTEM\CurrentControlSet\Services\MacFile\Parameters** subkey. Double-click on the subkey to expand it and display its values in the right-hand window.

4. Locate the **ServerOptions** value entry. Use the DWORD Editor to change the ServerOptions value by adding two to the current setting to enable clear-text passwords or subtracting two to disable them.

Allowing MacFile Server Local Password Saving

You can use the Registry to allow local password saving. To do this, follow these steps:

1. Launch Regedt32.

2. Select the Window menu option for **HKEY_LOCAL_MACHINE**.

3. Use the tree control in the left-hand window to navigate to the **SYSTEM\CurrentControlSet\Services\MacFile\Parameters** subkey. Double-click on the subkey to expand it and display its values in the right-hand window.

4. Locate the **ServerOptions** value entry. Use the DWORD Editor to change the ServerOptions value by adding 4 to the current setting to enable local password saving or subtracting 4 to disable it.

Setting MacFile Server Maximum Clients

You can use the Registry to maximize the number of users who can connect at any one time. To do this, follow these steps:

1. Launch Regedt32.

2. Select the Window menu option for **HKEY_LOCAL_MACHINE**.

3. Use the tree control in the left-hand window to navigate to the **SYSTEM\CurrentControlSet\Services\MacFile\Parameters\Volumes** subkey. Double-click on the subkey to expand it and display its values in the right-hand window.

4. Locate the **Microsoft UAM Volume** value entry. Use the Multi-String Editor to locate and change the **MaxUses** entry to the desired maximum client level.

TIP: *You can set the MaxUses value to **0xFFFFFFFF** to permit unlimited (infinite) users.*

9. Networking Protocols and Interoperability

Setting MacFile Server Volume Properties

You can use the Registry to set the properties of a given MacFile volume. To do this, follow these steps:

1. Launch Regedt32.

2. Select the Window menu option for **HKEY_LOCAL_MACHINE**.

3. Use the tree control in the left-hand window to navigate to the **SYSTEM\CurrentControlSet\Services\MacFile\Parameters\Volumes** subkey. Double-click on the subkey to select it and display its values in the right-hand window.

4. Locate the **Microsoft UAM Volume** value entry. Use the Multi-String Editor to locate and change the Properties entry using one of the values shown in Table 9.2.

Table 9.2   Setting MacFile volume properties.

Value	Definition
0000000000000000	Read/write access, no guests
0000000000000001	Read/write access, guests OK
1000000000000001	Read-only access, guests OK
1000000000000000 (the default)	Read-only access, no guests

Resetting MacFile Server User Passwords

You can use the Registry to clear an old or forgotten password so a user can enter a new one. To clear a password, follow these steps:

1. Launch Regedt32.

2. Select the Window menu option for **HKEY_LOCAL_MACHINE**.

3. Use the tree control in the left-hand window to navigate to the **SYSTEM\CurrentControlSet\Services\MacFile\Parameters\Volumes** subkey. Double-click on the subkey to expand it and display its values in the right-hand window.

4. Locate the **Microsoft UAM Volume** value entry. Use the Multi-String Editor to locate and change the Password entry to an empty value.

WARNING! You can clear a password, but you cannot change it because it is system-encrypted.

Fixing MacFile Server Volume Path Errors

You can use the Windows 2000 Registry to change the path to MacFile volumes to prevent errors after an administrator has moved the files. To do this, follow these steps:

1. Launch Regedt32.

2. Select the Window menu option for **HKEY_LOCAL_MACHINE**.

3. Use the tree control in the left-hand window to navigate to the **SYSTEM\CurrentControlSet\Services\MacFile\Parameters\Volumes** subkey. Double-click on the subkey to expand it and display its values in the right-hand window.

4. Locate the **Microsoft UAM Volume** value entry. Use the Multi-String Editor to locate and change the **Path** entry to the correct location.

Enabling the AppleTalk Router

Windows 2000 can also act as a router for the AppleTalk protocol by using the Registry to specify whether an AppleTalk router should start for a given computer. To do this, follow these steps:

1. Launch Regedt32.

2. Select the Window menu option for **HKEY_LOCAL_MACHINE**.

3. **Use the tree control in the left-hand window to navigate to the SYSTEM\CurrentControlSet\Services\Appletalk\ Parameters** subkey. Double-click on the subkey to expand it and display its values in the right-hand window.

4. Locate the **EnableRouter** value entry. Use the DWORD Editor to change the EnableRouter value to 1 to start the AppleTalk router for a machine or set EnableRouter to 0 if the router should start on another server.

Enabling Ethernet II Frames in NWLink

You can use the Registry to enable using Ethernet II Frames with IPX/SPX. To do this, follow these steps:

1. Launch Regedt32.

2. Select the Window menu option for **HKEY_LOCAL_MACHINE**.

9. Networking Protocols and Interoperability

3. Use the tree control in the left-hand window to navigate to the **SYSTEM\CurrentControlSet\Services\NWLinkIPX\NetConfig\ Drivers** subkey. Double-click on the subkey to expand it and display its values in the right-hand window.

4. Locate the **BindSAP** value entry. Use the DWORD Editor to change the BindSAP value to 1 to enable Ethernet II Frames or 0 to disable using them.

Setting NWLink Packet Types

You can use the Registry to set the packet type for the NWLink protocol. To do this, follow these steps:

1. Launch Regedt32.

2. Select the Window menu option for **HKEY_LOCAL_MACHINE**.

3. Use the tree control in the left-hand window to nπavigate to the SYSTEM\CurrentControlSet\Services\NWLinkIPX\NetConfig\ Drivers subkey. Double-click on the subkey to expand it and display its values in the right-hand window.

4. Locate the **PktType** value entry. Use the DWORD Editor to change the **PktType** value to one of the values shown in Table 9.3.

Table 9.3   Setting NWLink IPX/SPK packet types.

Value	Semantics
0	Ethernet II
1	Ethernet 802.5
2	802.2
3	SNAP
4	Arcnet

Setting NetBIOS **T1 Timeout** Values

You can use the Registry to set the **T1 timeout** value for packet retransmissions. This is useful if you are on a NetBIOS Frame (NBF) network with a server that has a very slow link. In this configuration, you should consider increasing this value. To do this, follow these steps:

1. Launch Regedt32.

2. Select the Window menu option for **HKEY_LOCAL_MACHINE**.

3. Use the tree control in the left-hand window to navigate to the **SYSTEM\CurrentControlSet\Services\NBF\Parameters** subkey. Double-click on the subkey to expand it and display its values in the right-hand window.

4. Locate the **DefaultT1Timeout** value entry. Use the DWORD Editor to change the DefaultT1Timeout to a higher value to compensate for slower networks.

Setting NetBIOS **T2 Timeout** Values

You can use the Registry to change the **T2 timeout** value for packet retransmissions. **T2** controls the time that NBF can wait after receiving an LLC (Logical Link Control) poll packet before responding. It must be much less than **T1**; one-half or less is a good general rule. To change the **T2 timeout** value, follow these steps:

1. Launch Regedt32.

2. Select the Window menu option for **HKEY_LOCAL_MACHINE**.

3. Use the tree control in the left-hand window to navigate to the **SYSTEM\CurrentControlSet\Services\NBF\Parameters** subkey. Double-click on the subkey to expand it and display its values in the right-hand window.

4. Locate the **DefaultT2Timeout** value entry. Use the DWORD Editor to change the DefaultT2Timeout value to a higher value to compensate for slower networks.

Setting NetBIOS **Ti Timeout** Values

You can use the Registry to change the **Ti timeout** (inactivity timer) value for packet retransmissions. When **Ti** expires, NBF sends an LLC poll packet to ensure that the link is still active. You should adjust this parameter only if NBF is connecting over networks with poor reliability characteristics, over slow networks, or to slow computers.

To change the **Ti timeout** value, follow these steps:

1. Launch Regedt32.

2. Select the Window menu option for **HKEY_LOCAL_MACHINE**.

9. Networking Protocols and Interoperability

3. Use the tree control in the left-hand window to navigate to the **SYSTEM\CurrentControlSet\Services\NBF\Parameters** subkey. Double-click on the subkey to expand it and display its values in the right-hand window.

4. Locate the **DefaultTiTimeout** value entry. Use the DWORD Editor to change the DefaultTiTimeout value to a higher value to compensate for slower networks.

Handling Unreliable Network Connections in NetBIOS

You can use the Registry to change the number of LLC I-frames sent before waiting for a remote response. To do this, follow these steps:

1. Launch Regedt32.

2. Select the Window menu option for **HKEY_LOCAL_MACHINE**.

3. Use the tree control in the left-hand window to navigate to the **SYSTEM\CurrentControlSet\Services\NBF\Parameters** subkey. Double-click on the subkey to expand it and display its values in the right-hand window.

4. Locate the **LLCMaxWindowSize** value entry. Use the DWORD Editor to change the LLCMaxWindowSize value to a relatively low value to handle unreliable network connections.

Fixing NBF Frame Maxouts in NetBIOS

The number of frames that a receiving computer is allowed to receive before sending an ACK to the sending computer is referred to as a receive window. In general, NBF has no receive window, unless it detects that the remote sending computer is running a version of IBM NetBEUI which does not support polling; in this case, NBF uses a receive window based on the value of **MaximumIncomingFrames** in the Registry. To adjust this value follow these steps:

1. Launch Regedt32.

2. Select the Window menu option for **HKEY_LOCAL_MACHINE**.

3. Use the tree control in the left-hand window to navigate to the **SYSTEM\CurrentControlSet\Services\NBF\Parameters** subkey. Double-click on the subkey to expand it and display its values in the right-hand window.

4. Locate the **MaximumIncomingFrames** value entry. Use the DWORD Editor to change the MaximumIncomingFrames value to a setting that is lower than the **MaximumOutgoingFrames** Registry value of the faster computer. This prevents swamping.

Fine-Tuning NBF Memory Usage in NetBIOS

You can modify the Registry to control how NBF works with available memory. To control NBF memory usage, follow these steps:

1. Launch Regedt32.

2. Select the Window menu option for **HKEY_LOCAL_MACHINE**.

3. Use the tree control in the left-hand window to navigate to the **SYSTEM\CurrentControlSet\Services\NBF\Parameters** subkey. Double-click on the subkey to expand it and display its values in the right-hand window.

4. Locate the **MaxRequests** value entry. Use the DWORD Editor to change the MaxRequests value to a value that maximizes the use of available NBF memory (this number is usually found by trial and error).

TIP: Setting the MaxRequests value to 0 disables memory usage optimization.

Fixing Slow Address Resolution Timeouts in NetBIOS

You can modify the Registry to fix timeouts over slow network connections due to address resolution failures. To do this, follow these steps:

1. Launch Regedt32.

2. Select the Window menu option for **HKEY_LOCAL_MACHINE**.

9. Networking Protocols and Interoperability

3. Use the tree control in the left-hand window to navigate to the **SYSTEM\CurrentControlSet\Services\NBF\Parameters** subkey. Double-click on the subkey to expand it and display its values in the right-hand window.

4. Locate the **AddNameQueryTimeout** value entry. Use the DWORD Editor to change the AddNameQueryTimeout value to a larger value to compensate for a slow network.

Fixing Slow Network Timeouts in NetBIOS

You can modify the Registry to fix timeouts over slow network connections. To do this, follow these steps:

1. Launch Regedt32.

2. Select the Window menu option for **HKEY_LOCAL_MACHINE**.

3. Use the tree control in the left-hand window to navigate to the **SYSTEM\CurrentControlSet\Services\NBF\Parameters** subkey. Double-click on the subkey to expand it and display its values in the right-hand window.

4. Locate the **GeneralTimeout** value entry. Use the DWORD Editor to change the GeneralTimeout value to a larger value to compensate for a slow network.

Fixing Name-Query Timeouts in NetBIOS

You can modify the Registry to fix timeouts over slow network connections due to name-query failures. To do this, follow these steps:

1. Launch Regedt32.

2. Select the Window menu option for **HKEY_LOCAL_MACHINE**.

3. Use the tree control in the left-hand window to navigate to the **SYSTEM\CurrentControlSet\Services\NBF\Parameters** subkey. Double-click on the subkey to expand it and display its values in the right-hand window.

4. Locate the **NameQueryTimeout** value entry. Use the DWORD Editor to change the NameQueryTimeout value to a larger value to compensate for a slow network.

Disabling Source Routing in NetBIOS

You can modify the Registry to disable source routing on token ring networks. To disable source routing on token ring networks, follow these steps:

1. Launch Regedt32.

2. Select the Window menu option for **HKEY_LOCAL_MACHINE**.

3. Use the tree control in the left-hand window to navigate to the **SYSTEM\CurrentControlSet\Services\NBF\Parameters** subkey. Double-click on the subkey to expand it and display its values in the right-hand window.

4. Locate the **QueryWithoutSourceRouting** value entry. Use the DWORD Editor to change the QueryWithoutSourceRouting value to 1 to turn off Source Routing on the token ring network.

Fixing DLC Timeouts

You can set several useful Registry parameters for DLC, including its timeouts. To set DLC timeouts, follow these steps:

1. Launch Regedt32.

2. Select the Window menu option for **HKEY_LOCAL_MACHINE**.

3. Use the tree control in the left-hand window to navigate to the **SYSTEM\CurrentControlSet\Services\DLC\Parameters\ [AdapterName]** subkey, where **[AdapterName]** is the network card communicating with the mainframe or HP printer. Double-click on the subkey to expand it and display its values in the right-hand window.

4. Locate the **T1TickOne, T1TickTwo, T2TickOne,** or **T2TickTwo** value entries. Use the DWORD Editor to change the appropriate value entry to a higher or lower value, as needed.

Enabling DLC Address Swapping

You can set several useful Registry parameters for DLC, including bit flipping Ethernet addresses. To enable bit flipping, follow these steps:

1. Launch Regedt32.

2. Select the Window menu option for **HKEY_LOCAL_MACHINE**.

9. Networking Protocols and Interoperability

3. Use the tree control in the left-hand window to navigate to the
SYSTEM\CurrentControlSet\Services\DLC\Parameters
[AdapterName] subkey, where *[AdapterName]* is the
network card communicating with the mainframe or HP
printer. Double-click on the subkey to expand it and display its
values in the right-hand window.

4. Locate the **Swap** value entry. Use the DWORD Editor to change
the Swap value to 1 to enable bit flipping and 0 to disable it.

Local and Networked Printers

In Brief

Printing, particularly networked printing, is one of the major functions of an operating system. Windows 2000 provides excellent support in this area, including GUI (graphical user interface) support for installing printers, assigning printer ports, and managing printer properties.

Installing Printers

Although printers are commonly taken to be physical objects and are installed as hardware, in Windows 2000 they can be viewed as software elements called *print queues*. Windows 2000 provides a powerful graphical wizard to handle adding a new physical printer and creating a print queue for it.

Assigning Printer Ports

Ports in Windows 2000 are no trivial matter. Aside from the usual parallel and serial ports physically attached to a given machine, there are also shared network printers, Named Pipes, fax systems, JET-Direct ports for Hewlett-Packard DLC (Data Link Control) network drivers, and even Internet printing to IP (Internet Protocol) addresses and LPD (Line Printer Daemon) Unix servers. Each device has either an installation program or documentation to guide administrators through the process of assigning its printer port values. If for some reason they need to be changed using the techniques outlined later in this chapter, be sure to consult the documentation for the device prior to making changes.

Managing Printer Properties

Windows 2000 has a very helpful GUI that you can use to manage the properties of its printers. The GUI allows you to add and delete printers and printer-like devices, as well as set the specialized properties of an individual printer. When the GUI can't handle a given situation, however, it's time to dive into the Registry.

Immediate Solutions

Enabling Printer Alert Sounds

When a remote print job returns an error on a print server, Windows 2000 permits enabling a beep sound to alert the user each time the job is retried (every 10 seconds). To enable the beeping event, follow these steps:

1. Launch Regedt32.

2. Select the Window menu option for **HKEY_LOCAL_MACHINE**.

3. Use the tree control in the left-hand window to navigate to the **SYSTEM\CurrentControlSet\Control\Print** subkey. Double-click on the subkey to expand it and display its values in the right-hand window.

4. Locate the **BeepEnabled** value entry. Use the DWORD Editor to change the BeepEnabled value to 1 to enable beeping.

Disabling a Printer Browser Thread

Computers running Windows 2000 have a feature called a Browser that advertises the existence of resources to other computers on the network. To disable this feature on a given Windows 2000 computer, tweak the Registry as follows:

1. Launch Regedt32.

2. Select the Window menu option for **HKEY_LOCAL_MACHINE**.

3. Use the tree control in the left-hand window to navigate to the **SYSTEM\CurrentControlSet\Control\Print** subkey. Double-click on the subkey to expand it and display its values in the right-hand window.

4. Locate the **DisableServerThread** value entry. If the value does not exist, you can add it. Use the DWORD Editor to change the DisableServerThread value to 1 to stop advertising printers on the computer.

Slowing Down Fast Printing

Fast Printing is a Windows 2000 feature that allows print jobs to be sent to a printer while they are still being created by their originating application. In some cases (such as when a network becomes loaded), this can result in too much data being sent to the printer. To set a value that, when reached, slows down the rate of data throughput, tweak the Registry as follows:

1. Launch Regedt32.

2. Select the Window menu option for **HKEY_LOCAL_MACHINE**.

3. Use the tree control in the left-hand window to navigate to the **SYSTEM\CurrentControlSet\Control\Print** subkey. Double-click on the subkey to expand it and display its values in the right-hand window.

4. Locate the **FastPrintSlowDownThreshold** value entry. If the value does not exist, you can create it. Use the DWORD Editor to change the FastPrintSlowDownThreshold value to a lower setting (in milliseconds). Lowering this setting, causes Fast Printing to be more sensitive to busy networks.

Setting the Speed of Fast Printing

In some cases, the Fast Printing capability in Windows 2000 can result in a printer shutting down because the application pauses when sending data. To correct this, tweak the Registry as follows:

1. Launch Regedt32.

2. Select the Window menu option for **HKEY_LOCAL_MACHINE**.

3. Use the tree control in the left-hand window to navigate to the **SYSTEM\CurrentControlSet\Control\Print** subkey. Double-click on the subkey to expand it and display its values in the right-hand window.

4. Locate the **FastPrintThrottleTimeout** value entry. If the value does not exist, you can create it. Use the DWORD Editor to change the FastPrintThrottleTimeout value to a lower value so that available data is sent more slowly (the slowest rate is one byte per frame). This setting is in milliseconds.

Setting the Fast Printing **Timeout** Value

Another related setting in the Windows 2000 Registry is
FastPrintWaitTimeout. This setting also allows Fast Printing behavior to handle applications that pause when printing data. To
adjust the amount of time Fast Printing waits, tweak the Registry
as follows:

1. Launch Regedt32.

2. Select the Window menu option for **HKEY_LOCAL_MACHINE**.

3. Use the tree control in the left-hand window to navigate to the
   SYSTEM\CurrentControlSet\Control\Print subkey. Double-
   click on the subkey to expand it and display its values in the
   right-hand window.

4. Locate the **FastPrintWaitTimeout** value entry. If the value
   does not exist, you can create it. Use the DWORD Editor to
   change the FastPrintWaitTimeout value to a higher value to
   handle the slow application.

Setting the Print Queue Decay Time

On busy networks, which printers are available for printing at any
given time can change quickly. To help keep users from continually
having to reselect printers because the one they initially chose is unavailable, you can set a *decay time* for the list of printers (print queues).
This can help clean up the list of printers so only those that are truly
available appear. To set the decay time, tweak the Registry as follows:

1. Launch Regedt32.

2. Select the Window menu option for **HKEY_LOCAL_MACHINE**.

3. Use the tree control in the left-hand window to navigate to the
   SYSTEM\CurrentControlSet\Control\Print subkey. Double-
   click on the subkey to expand it and display its values in the
   right-hand window.

4. Locate the **NetPrinterDecayPeriod** value entry. If the value
   does not exist, you can create it. Use the DWORD Editor to
   change the NetPrinterDecayPeriod value to a lower value so
   that the list in the Browse dialog box is refreshed over the
   network more often.

WARNING! *The NetPrinterDecayPeriod value is, for some odd reason, specified in milliseconds—one hour is 3,600,000 milliseconds. Mistakenly assuming that the value is specified in minutes can tie up a network with continual browser refresh requests.*

Controlling the Printer Port Thread's Priority

Each printer port has one or more threads that are used to send output to physical printers. You can control the priority of these threads by changing the Registry as follows:

1. Launch Regedt32.

2. Select the Window menu option for **HKEY_LOCAL_MACHINE**.

3. Use the tree control in the left-hand window to navigate to the **SYSTEM\CurrentControlSet\Control\Print** subkey. Double-click on the subkey to expand it and display its values in the right-hand window.

4. Locate the **PortThreadPriority** value entry. If the value does not exist, you can create it. Use the DWORD Editor to change the PortThreadPriority value to one of the values shown in Table 10.1.

Table 10.1   Thread priority levels.

Value	Description
0	Normal thread priority
1	Above normal thread priority
0xFFFFFFFF	Below normal thread priority

Setting a Printer's Scheduling Thread Priority

The task of assigning print jobs involves a prioritized thread. If you observe uncharacteristic delays in scheduling print jobs, you might need to adjust the Registry. You can adjust print performance and the priority used to assign print jobs to ports by setting the **SchedulerThreadPriority** setting. You can also use this entry to improve overall system performance by reducing the priority of the Scheduler thread. Here is how you make the change:

1. Launch Regedt32.

2. Select the Window menu option for **HKEY_LOCAL_MACHINE**.

3. Use the tree control in the left-hand window to navigate to the **SYSTEM\CurrentControlSet\Control\Print** subkey. Double-click on the subkey to expand it and display its values in the right-hand window.

4. Locate the **SchedulerThreadPriority** value entry. If the value does not exist, you can create it. Use the DWORD Editor to change the SchedulerThreadPriority value to one of the values shown in Table 10.1.

Setting a Printer's Priority Class

On older NT systems that might be part of a Windows 2000 server-based network, applications and administrators set printer priority classes by tweaking the Registry as follows:

1. Launch Regedt32.

2. Select the Window menu option for **HKEY_LOCAL_MACHINE**.

3. Use the tree control in the left-hand window to navigate to the **SYSTEM\CurrentControlSet\Control\Print** subkey. Double-click on the subkey to expand it and display its values in the right-hand window.

4. Locate the **PriorityClass** value entry. If the value does not exist, you can create it. Use the DWORD Editor to change the PriorityClass value to a value shown in Table 10.2.

WARNING! In Windows 2000, the PriorityClass entry is superseded by the SpoolerPriority entry; setting PriorityClass has no effect in Windows 2000, only in older versions of NT.

Table 10.2   Class priority levels.

Value	Description
0	Normal priority class
1	High priority class
0xFFFFFFFF	Idle priority class

Setting a Printer's Spooler Priority

You can assign a priority class to the threads for the print spooler by adding the **SpoolerPriority** Registry entry and setting its value. To do this, follow these steps:

1. Launch Regedt32.

2. Select the Window menu option for **HKEY_LOCAL_MACHINE**.

3. Use the tree control in the left-hand window to navigate to the **SYSTEM\CurrentControlSet\Control\Print** subkey. Double-click on the subkey to expand it and display its values in the right-hand window.

4. Create a **SpoolerPriority** entry, of type REG_DWORD. Choose a SpoolerPriority value from one of the values shown earlier in this chapter in Table 10.2.

Setting the Path to a Printer's Configuration DLL

Windows 2000 does not really configure printers. Instead, a printer manufacturer supplies a dynamic link library (DLL) that is responsible for providing a dialog box and the internal hooks to handle the work. If a given printer cannot locate its configuration DLL, an error occurs when users try to configure the printer. To fix this, tweak the Registry as follows:

1. Launch Regedt32.

2. Select the Window menu option for **HKEY_LOCAL_MACHINE**.

3. Use the tree control in the left-hand window to navigate to the **SYSTEM\CurrentControlSet\Control\Print\Environments\[NT Environment]\Print Processors\[Printer Driver Name]** subkey, where *[NT Environment]* is a value describing the processor (such as "Windows NT x86"), and *[Printer Driver Name]* is a string describing the printer (such as "Epson Stylus Color"). Double-click on the subkey to expand it and display its values in the right-hand window.

4. Locate the **ConfigurationFile** value entry. Use the String Editor to change the ConfigurationFile value to the correct path for the configuration DLL for a given printer (from documentation or a file search).

Setting the Path to a Printer's Data File

If a given printer cannot locate its configuration data file, an error occurs when users try to configure the printer. This information is usually put in the Registry by an installation program, but the file can later be accidentally moved, renamed, or deleted. To change the Registry to reflect the current location and/or name of the file, tweak the Registry as follows:

1. Launch Regedt32.

2. Select the Window menu option for **HKEY_LOCAL_MACHINE**.

3. Use the tree control in the left-hand window to navigate to the **SYSTEM\CurrentControlSet\Control\Print\Environments\[NT Environment]\Print Processors\[*Printer Driver Name*]** subkey, where *[NT Environment]* is a value describing the processor (such as "Windows NT x86"), and *[Printer Driver Name]* is a string describing the printer (such as "Epson Stylus Color"). Double-click on the subkey to expand it and display its values in the right-hand window.

4. Locate the **DataFile** value entry. Use the String Editor to change the DataFile value to the correct value for the configuration data file (from documentation after reinstalling or from a file search).

Related solution:	Found on page:
Getting the Installed Windows 2000 System Path Information	146

Setting The DLL for a Printer's Driver

The driver file for a printer in Windows 2000 must reside at the location entered in the Registry. This is yet another entry that must be correct - sometimes, this entry is made incorrectly or the file gets trashed or moved. If you can't get a printer to work, use the following steps to check the Registry to make sure the driver path is valid:

1. Launch Regedt32.

2. Select the Window menu option for **HKEY_LOCAL_MACHINE**.

3. Use the tree control in the left-hand window to navigate to the **SYSTEM\CurrentControlSet\Control\Print\Environments\[NT Environment]\Print Processors\[*Printer Driver Name*]** subkey, where *[NT Environment]* is a value describing the

processor (such as "Windows NT x86"), and *[Printer Driver Name]* is a string describing the printer (such as "Epson Stylus Color"). Double-click on the subkey to expand it and display its values in the right-hand window.

4. Locate the **Driver** value entry. Use the String Editor to make sure the Driver value points to the actual location of the proper driver DLL and that a valid DLL resides at that location.

Determining the Version of a Printer's Driver

Although most printer drivers include information in the Printer Properties area, including the current version number, some don't. You can use the Registry to determine the version of a printer's driver by following these steps:

1. Launch Regedt32.

2. Select the Window menu option for **HKEY_LOCAL_MACHINE**.

3. Use the tree control in the left-hand window to navigate to the **SYSTEM\CurrentControlSet\Control\Print\Environments\[NT Environment]\Print Processors\[Printer Driver Name]** subkey, where *[NT Environment]* is a value describing the processor (such as "Windows NT x86"), and *[Printer Driver Name]* is a string describing the printer (such as "Epson Stylus Color"). Double-click on the subkey to expand it and display its values in the right-hand window.

4. Locate the **Version** value entry. Use the DWORD Editor to view the current version number.

Related solution:	Found on page:
Getting the Installed Windows 2000 Version Information	145

Setting the Print Monitor's Driver

Each network printer port has a printer monitor to run it. If you experience problems with a network port, you can use the following steps to check the Registry to ensure that a network port's driver DLL is in the right place:

1. Launch Regedt32.

2. Select the Window menu option for **HKEY_LOCAL_MACHINE**.

3. Use the tree control in the left-hand window to navigate to the **SYSTEM\CurrentControlSet\Control\Print\Monitors\[*Provider Network Port]* subkey, where *[Provider Network Port]* is the port name for the network printer port. Double-click on the subkey to expand it and display its values in the right-hand window.

4. Locate the **Driver** value entry. Use the String Editor to ensure that the path and DLL are valid.

Setting the Print Switch for a Given Port

Each printer port is sent a format control character by the LPR (line printer remote) print monitor. In rare circumstances, such as when this value has been lost or corrupted, you might need to tweak the Registry to change the default format control character. To do this, follow these steps:

1. Launch Regedt32.

2. Select the Window menu option for **HKEY_LOCAL_MACHINE**.

3. Use the tree control in the left-hand window to navigate to the **SYSTEM\CurrentControlSet\Control\Print\Monitors\LocalPort\ Ports\[*Port Name]* subkey, where *[Port Name]* is the printer port (such as "LPT1:"). Double-click on the subkey to expand it and display its values in the right-hand window.

4. Locate the **PrintSwitch** value entry. Use the String Editor to set PrintSwitch to the correct character (normally l, but in some cases f, as outlined in the printer documentation).

Examining Individual Printer Settings

The Registry contains considerable information about each printer on a Windows 2000 network. Here's how to use the Registry to check a printer's settings when troubleshooting:

1. Launch Regedt32.

2. Select the Window menu option for **HKEY_LOCAL_MACHINE**.

3. Use the tree control in the left-hand window to navigate to the **SYSTEM\CurrentControlSet\Control\Print\Printers\[*Printer*

Name] subkey, where *[Printer Name]* is the value of the Printer's Print Queue in the Printers folder (such as "Epson Stylus Color"). Double-click on the subkey to expand it and display its values in the right-hand window.

4. Values vary by printer. Check for a value that matches the problem you're attempting to fix. The information should be in the printer documentation.

Setting the Print Provider DLL

Each installed print provider has a Registry entry pointing to its DLL. Here's how to access that entry if problems occur:

1. Launch Regedt32.

2. Select the Window menu option for **HKEY_LOCAL_MACHINE**.

3. Use the tree control in the left-hand window to navigate to the **SYSTEM\CurrentControlSet\Control\Print\Providers***[Print Services Name]* subkey, where *[Print Services Name]* is the provider name. Double-click on the subkey to expand it and display its values in the right-hand window.

4. Locate the **Name** value entry. Use the String Editor to make sure the Name value points to the correct path and DLL (as indicated in the documentation or from a file search).

Controlling Whether Printer Errors Pop up on Remote Machines

You can control whether printer errors or print completion messages show pop-up dialog boxes on remote machines. To do so, tweak the Registry for a given print provider as follows:

1. Launch Regedt32.

2. Select the Window menu option for **HKEY_LOCAL_MACHINE**.

3. Use the tree control in the left-hand window to navigate to the **SYSTEM\CurrentControlSet\Control\Print\Providers***[Print Services Name]* subkey, where *[Print Services Name]* is the provider in question. Double-click on the subkey to expand it and display its values in the right-hand window.

4. If necessary, add a **NetPopup** value entry, of type REG_DWORD. Set its value to 1 to enable remote error pop-up dialog boxes or 0 to disable them.

Enabling Trusted Printing

To prevent unauthorized printer servers from gaining control of a secure printer resource (such as a Phaser II color printer, where pages cost several dollars each), you can require all print jobs to come from a trusted server. To do this, follow these steps:

1. Launch Regedt32.

2. Select the Window menu option for **HKEY_LOCAL_MACHINE**.

3. Use the tree control in the left-hand window to navigate to the **SYSTEM\CurrentControlSet\Control\Print\Providers\LanMan Print Services** subkey. Double-click on the subkey to expand it and display its values in the right-hand window.

4. Locate the **LoadTrustedDrivers** value entry. Use the DWORD Editor to change the LoadTrustedDrivers value to 1 to enable only trusted printer servers.

Changing the Print Spool Directory

By default, Windows 2000 places print spooler activities on the system disk. This might be fine in your installation; on the other hand, you might want to move these activities to a volume with more free space. To do so, tweak the Registry as follows:

1. Launch Regedt32.

2. Select the Window menu option for **HKEY_LOCAL_MACHINE**.

3. Use the tree control in the left-hand window to navigate to the **SOFTWARE\Microsoft\ Windows NT\CurrentVersion\Print\ Printers** subkey. Double-click on the subkey to expand it and display its values in the right-hand window.

4. Locate the **DefaultSpoolDirectory** value entry. Use the String Editor to modify the path information.

Stopping Print Job Logging to the Event Log

You might have noticed that Windows 2000 logs print jobs to the application event log on the print server. If you do not need these events logged, why cause the additional system overhead and also have these logs filling up with this information? Here is a handy Registry tweak to stop this logging behavior:

1. Launch Regedt32.

2. Select the Window menu option for **HKEY_LOCAL_MACHINE**.

3. Use the tree control in the left-hand window to navigate to the **SYSTEM\CurrentControlSet\Control\Print\Providers**.

4. Locate the **EventLog** value entry. Set this value to 0 to disable logging.

Chapter 11

Routing and Remote Access Service

In Brief

Routing and Remote Access Service (RRAS) provides remote login capabilities via modems, ISDN connectors, and X.25 digital WANs to an entire Windows 2000-based LAN or WAN. It can serve the needs of traveling executives and sales personnel, provide complete Internet Service Provider capabilities via Transmission Control Protocol/Internet Protocol (TCP/IP), and create fully functional wide area networks over phone or dedicated connections. RRAS features include gateways for many types of network protocols, such as Network Basic Input-Output System (NetBIOS), Mac, Internet Packet Exchange (IPX), Routing Information Procol (RIP), and Network Device Interface Specification (NDIS)), a group of General Services, auditing capabilities, and TCP/IP and Point-to-Point Protocol (PPP) support.

Protocol Support

RRAS provides dialup support for a number of networking protocols, including the following:

- NetBIOS
- AsyncMac
- NdisWAN
- NwLnkIPX
- NwLnkRIP
- Rdr
- TCP/IP and PPP

RRAS does the work of translating networking protocols between the dialup-based computer and the networked computer on the LAN or intranet. It also makes sure security issues are handled correctly and prevents "spoofing" so that remote clients cannot use their access to bypass normal verification procedures.

Auditing

There is no auditing by default in Windows 2000 RRAS; the administrator has to enable RRAS by configuring the Registry. This chapter describes how to modify the Registry to enable RRAS auditing.

General Windows 2000 Services

The general services provided by Windows 2000 RRAS include authentication and retrying authentication, authentication timeouts, automatic disconnects for idle connections, callbacks for remote connections, and enabling the NetBIOS gateway. Each of these features has a useful Registry entry that is covered in an Immediate Solution in this chapter.

TCP/IP and PPP

RRAS support for TCP/IP and PPP includes support for the following:

- *Windows Internet Name Service (WINS)*—RRAS provides full support for WINS name servers, including backups and multiple servers.

- *Address Resolution Protocol (ARP)*—RRAS supports ARP outgoing multicast packets, but does not support incoming multicasts or broadcasts.

- *Compression Control Protocol (COMPCP)*—RRAS supports COMPCP, a way to employ Lempel-Zev compression on PPP data streams. RRAS supports COMPCP by providing a connection to a COMPCP dynamic link library (DLL), but the DLL itself must be provided by a third party.

Immediate Solutions

Enabling RRAS Auditing

By default, RRAS sessions are not logged (audited). To turn on saving all RRAS session information to the Windows 2000 Event Log file, modify the Registry as follows:

1. Launch Regedt32.

2. Select the Window menu option for **HKEY_LOCAL_MACHINE**.

3. Use the tree control in the left-hand window to navigate to the **SYSTEM\CurrentControlSet\Services\RemoteAccess\ Parameters** subkey. Double-click on the subkey to expand it and display its values in the right-hand window.

4. Locate the **EnableAudit** value entry. If the value does not exist, you can create it. Use the DWORD Editor to change the EnableAudit value to 1 to enable RRAS auditing.

Setting RRAS Port Logging

By default, RRAS connections to serial ports are not logged (audited). To enable saving all RRAS serial connection information to the Windows 2000 Device Log file (\%systemroot%\system32\RRAS\device.log), modify the Registry as follows:

1. Launch Regedt32.

2. Select the Window menu option for **HKEY_LOCAL_MACHINE**.

3. Use the tree control in the left-hand window to navigate to the **SYSTEM\CurrentControlSet\Services\RRASman\ Parameters** subkey. Double-click on the subkey to expand it and display its values in the right-hand window.

4. Locate the **Logging** value entry. If the value does not exist, you can create it. Use the DWORD Editor to change the Logging value to 1 to enable RRAS logging.

Changing the Maximum Allowed RRAS Authentication Retries

One of several key RRAS entries in Windows 2000's general Registry section controls the maximum allowable retries on a RRAS connection. This prevents the system from mistaking connection difficulties for a security problem. To raise the maximum allowable retries on a RRAS connection, change the Registry as follows:

1. Launch Regedt32.

2. Select the Window menu option for **HKEY_LOCAL_MACHINE**.

3. Use the tree control in the left-hand window to navigate to the **SYSTEM\CurrentControlSet\Services\RemoteAccess\ Parameters** subkey. Double-click the subkey to expand it and display its values in the right-hand window.

4. Locate the **AuthenticateRetries** value entry. If the value does not exist, you can add it. Set its REG_DWORD value to a value appropriate for your network.

Changing the RRAS Authentication Timeout

A key RRAS entry in Windows 2000's general Registry section allows you to set a time limit during which users must successfully authenticate themselves before the connection is closed. To configure this feature, follow these steps:

1. Launch Regedt32.

2. Select the Window menu option for **HKEY_LOCAL_MACHINE**.

3. Use the tree control in the left-hand window to navigate to the **SYSTEM\CurrentControlSet\Services\RemoteAccess\ Parameters** subkey. Double-click on the subkey to expand it and display its values in the right-hand window.

4. Locate the **AuthenticateTime** value entry. If the value does not exist, you can create it. Set its REG_DWORD value to a value appropriate for your network.

Setting the Inactive RRAS Connection Timeout

Another key RRAS entry in Windows 2000's general Registry section enables you to set the time allowed for an inactive connection before the system automatically disconnects. To configure this timeout setting, follow these steps:

1. Launch Regedt32.

2. Select the Window menu option for **HKEY_LOCAL_MACHINE**.

3. Use the tree control in the left-hand window to navigate to the **SYSTEM\CurrentControlSet\Services\RemoteAccess\ Parameters** subkey. Double-click on the subkey to expand it and display its values in the right-hand window.

4. Locate the **AutoDisconnect** value entry. If the value does not exist, you can create it. Set its REG_DWORD value to the time in minutes you wish to allow inactive connections to remain alive.

WARNING! **Setting the AutoDisconnect value to 0 turns off AutoDisconnect.**

Changing the RRAS Callback Wait Time

Another key RRAS entry in Windows 2000's general Registry section controls the amount of time the server waits before calling a client back via the callback feature. To set this feature, follow these steps:

1. Launch Regedt32.

2. Select the Window menu option for **HKEY_LOCAL_MACHINE**.

3. Use the tree control in the left-hand window to navigate to the **SYSTEM\CurrentControlSet\Services\RemoteAccess\ Parameters** subkey. Double-click on the subkey to expand it and display its values in the right-hand window.

4. Locate the **CallbackTime** value entry. If the value does not exist, you can create it. Set its REG_DWORD value to the value in seconds that is most appropriate for your network.

Setting RRAS AsyncMac Frame Sizes

A noisy data line for RRAS dialup can cause dropouts with large frame sizes, in which case you need to reduce the frame size. Here's how to adjust the Registry to reduce the frame size:

1. Launch Regedt32.

2. Select the Window menu option for **HKEY_LOCAL_MACHINE**.

3. Use the tree control in the left-hand window to navigate to the **SYSTEM\CurrentControlSet\Services\AsyncMac\ Parameters** subkey. Double-click on the subkey to expand it and display its values in the right-hand window.

4. Locate the **MaxFrameSize** value entry. You can add this value if it does not exist. Use the DWORD Editor to change the MaxFrameSize value to a lower setting to reduce dropouts on noisy lines.

Setting RRAS NetBIOS Gateway Timeouts

With slow connections or noisy lines, RRAS might have problems with NetBIOS gateway timeouts and need increased wait time. Here's how to adjust the Registry to configure the gateway timeout setting:

1. Launch Regedt32.

2. Select the Window menu option for **HKEY_LOCAL_MACHINE**.

3. Use the tree control in the left-hand window to navigate to the **SYSTEM\CurrentControlSet\Services\AsyncMac\ Parameters** subkey. Double-click on the subkey to expand it and display its values in the right-hand window.

4. Locate the **TimeoutBase** value entry. If the value does not exist, you can add it. Use the DWORD Editor to change the TimeoutBase value to a higher setting in seconds to reduce timeouts on noisy lines.

Setting the RRAS NDIS IEEE Address

You might need to reset the initial address bytes of NDIS information. Here's the Registry value to alter in order to accomplish this:

1. Launch Regedt32.

2. Select the Window menu option for **HKEY_LOCAL_MACHINE**.

3. Use the tree control in the left-hand window to navigate to the **SYSTEM\CurrentControlSet\Services\NdisWan\Parameters** subkey. Double-click on the subkey to expand it and display its values in the right-hand window.

4. Locate the **NetworkAddress** value entry. If the value does not exist you can add it. Use the String Editor to change the NetworkAddress value to the value your system needs according to the documentation.

Disabling RRAS Dialin for NetBIOS

You can turn off NetBIOS RRAS dialups when necessary by directly accessing the Registry. Here's how to change the Registry to turn off NetBIOS RRAS dialups:

1. Launch Regedt32.

2. Select the Window menu option for **HKEY_LOCAL_MACHINE**.

3. Use the tree control in the left-hand window to navigate to the **SYSTEM\CurrentControlSet\Services\NwlnkIpx\Parameters** subkey. Double-click on the subkey to expand it and display its values in the right-hand window.

4. Locate the **DisableDialinNetbios** value entry. If the value is not present, you can create it. Use the DWORD Editor to change the DisableDialinNetbios value to 1 to disable NetBIOS logins.

Controlling RRAS NetBIOS Broadcast Packet Routing

You can use the Registry to control how RRAS forwards NetBIOS broadcast packets to the network. To do this, follow these steps:

1. Launch Regedt32.

2. Select the Window menu option for **HKEY_LOCAL_MACHINE**.

3. Use the tree control in the left-hand window to navigate to the **SYSTEM\CurrentControlSet\Services\NwlnkRip\Parameters**

subkey. Double-click on the subkey to expand it and display its values in the right-hand window.

4. Locate the **NetbiosRouting** value entry. If the value does not exist you can create it. Use the DWORD Editor to change the NetbiosRouting value to one of the values shown in Table 11.1.

Table 11.1   Setting NetBIOS packet routing.

Value	Packet Routing Level
0	No forwarding
2	Client to WAN
4	WAN to client
6	WAN to client, and client to WAN

Forcing RRAS Serial FIFO Queuing

FIFO (first in, first out) queuing on serial ports is more efficient than LIFO (last in, first out), but you must enable it at the Registry level. Here's how to configure the Registry to enable FIFO queuing:

1. Launch Regedt32.

2. Select the Window menu option for **HKEY_LOCAL_MACHINE**.

3. Use the tree control in the left-hand window to navigate to the **SYSTEM\CurrentControlSet\Services\Serial** subkey. Double-click on the subkey to expand it and display its values in the right-hand window.

4. Locate the **ForceFifoEnable** value entry. Use the DWORD Editor to change the ForceFifoEnable value to 1 to force the serial port to use FIFO queuing.

Enabling RRAS Serial FIFO Logging

After you enable FIFO queuing for serial ports, you can enable it via the Registry to be logged and catch problems. To enable RRAS serial FIFO logging, follow these steps:

1. Launch Regedt32.

2. Select the Window menu option for **HKEY_LOCAL_MACHINE**.

3. Use the tree control in the left-hand window to navigate to the **SYSTEM\CurrentControlSet\Services\Serial** subkey. Double-click on the subkey to expand it and display its values in the right-hand window.

4. Locate the **LogFifo** value entry. Use the DWORD Editor to change the LogFifo value to 1 to enable FIFO logging for that port.

Permitting RRAS Serial Port Sharing

A single serial port can have more than one RRAS session on it, provided you configure the Registry correctly. To do this, follow these steps:

1. Launch Regedt32.

2. Select the Window menu option for **HKEY_LOCAL_MACHINE**.

3. Use the tree control in the left-hand window to navigate to the **SYSTEM\CurrentControlSet\Services\Serial** subkey. Double-click on the subkey to expand it and display its values in the right-hand window.

4. Locate the **PermitShare** value entry. Use the DWORD Editor to change the PermitShare value to 1 to enable RRAS serial port sharing.

Setting the RRAS **TxFIFO** Queue Size

FIFO queuing on serial ports is more efficient than LIFO. When the system is using FIFO, you must set the queue size at the Registry level, as follows:

1. Launch Regedt32.

2. Select the Window menu option for **HKEY_LOCAL_MACHINE**.

3. Use the tree control in the left-hand window to navigate to the **SYSTEM\CurrentControlSet\Services\Serial** subkey. Double-click on the subkey to expand it and display its values in the right-hand window.

4. Locate the **TxFIFO** value entry. Use the DWORD Editor to change the TxFIFO value to the maximum TCP/IP queued sessions on the port that you want the system to allow.

Setting the RRAS **RxFIFO** Queue Size

FIFO queuing on serial ports is more efficient than LIFO. When the system is using FIFO, you must set the queue size at the Registry level, as follows:

1. Launch Regedt32.

2. Select the Window menu option for **HKEY_LOCAL_MACHINE**.

3. Use the tree control in the left-hand window to navigate to the **SYSTEM\CurrentControlSet\Services\Serial** subkey. Double-click on the subkey to expand it and display its values in the right-hand window.

4. Locate the **RxFIFO** value entry. Use the DWORD Editor to change the RxFIFO value to the maximum NetBIOS queued sessions on the port the system should allow.

Changing the RRAS WINS Server Name

You can use the Registry to set a number of RRAS parameters that involve TCP/IP. One Registry setting enables you to change the WINS server's IP name that remote access clients use. To change the WINS server's IP name, follow these steps:

1. Launch Regedt32.

2. Select the Window menu option for **HKEY_LOCAL_MACHINE**.

3. Use the tree control in the left-hand window to navigate to the **SYSTEM\CurrentControlSet\Services\RemoteAccess\ Parameters\IP** subkey. Double-click on the subkey to expand it and display its values in the right-hand window.

4. Locate the **WINSNameServer** value entry. If the values does not exist, you can add it. Use the String Editor to change the WINSNameServer value to the desired IP name or address for the WINS server for the network.

Setting the RRAS WINS Server Backup

Another Registry setting allows you to specify the WINS server's backup server address that remote access clients use. To do this, follow these steps:

1. Launch Regedt32.

2. Select the Window menu option for **HKEY_LOCAL_MACHINE**.

3. Use the tree control in the left-hand window to navigate to the **SYSTEM\CurrentControlSet\Services\RemoteAccess\ Parameters\IP** subkey. Double-click on the subkey to expand it and display its values in the right-hand window.

4. Locate the **WINSNameServerBackup** value entry. If the value does not exist, you can add it. Use the String Editor to change the WINSNameServerBackup value to the desired IP name or address for the network's WINS backup server.

Setting PPP RRAS Terminate-Ack Timeouts

When RRAS uses PPP, it needs guidance on how to configure its networking requests based on the behavior of the local system. These settings are found in the Registry. One Registry setting controls the number of terminate-ack messages that are sent before the network assumes a client has lost its connection. Here's how to configure the Registry setting:

1. Launch Regedt32.

2. Select the Window menu option for **HKEY_LOCAL_MACHINE**.

3. Use the tree control in the left-hand window to navigate to the **SYSTEM\CurrentControlSet\Services\Rasman\PPP** subkey. Double-click on the subkey to expand it and display its values in the right-hand window.

4. Locate the **MaxTerminate** value entry. Use the DWORD Editor to change the MaxTerminate value to a setting that meets the needs of the current network environment. A higher setting (such as 10) results in a greater likelihood of acknowledging a connection termination (but with the unavoidable network performance loss). A low setting (such as 1) can result in not acknowledging a terminated RRAS connection, with the resulting loss of system resources to a dead connection.

Setting PPP RRAS Configure-Request Timeouts

Another Registry setting for RRAS and PPP controls the number of configure-request messages to send before the network assumes a client has lost its connection. Here's how to configure this Registry setting:

1. Launch Regedt32.

2. Select the Window menu option for **HKEY_LOCAL_MACHINE**.

3. Use the tree control in the left-hand window to navigate to the **SYSTEM\CurrentControlSet\Services\Rasman\PPP** subkey. Double-click on the subkey to expand it and display its values in the right-hand window.

4. Locate the **MaxConfigure** value entry. Use the DWORD Editor to change the MaxConfigure value to a setting that meets the needs of the current network environment.

Setting PPP RRAS Configure-Nak Timeouts

Another Registry setting regarding RRAS and PPP controls the number of configure-nak messages to send before the network assumes a client has lost its connection. Here's how to configure this Registry setting:

1. Launch Regedt32.

2. Select the Window menu option for **HKEY_LOCAL_MACHINE**.

3. Use the tree control in the left-hand window to navigate to the **SYSTEM\CurrentControlSet\Services\Rasman\PPP** subkey. Double-click on the subkey to expand it and display its values in the right-hand window.

4. Locate the **MaxFailure** value entry. Use the DWORD Editor to change the MaxFailure value to a setting that meets the needs of the current network environment.

Setting PPP RRAS Configure-Reject Timeouts

Another Registry setting related to RRAS and PPP controls the number of configure-reject messages to send before the network assumes a client has lost its connection. Here's how to configure this Registry setting:

1. Launch Regedt32.

2. Select the Window menu option for **HKEY_LOCAL_MACHINE**.

3. Use the tree control in the left-hand window to navigate to the **SYSTEM\CurrentControlSet\Services\Rasman\PPP** subkey. Double-click on the subkey to expand it and display its values in the right-hand window.

4. Locate the **MaxReject** value entry. Use the DWORD Editor to change the MaxReject value to a setting that meets the needs of the current network environment.

Setting PPP RRAS Packet Resend Wait Times

One more Registry setting related to RRAS and PPP controls the time RRAS waits before it assumes a packet is lost. The value is used to configure a Windows timer that, if allowed to trigger, causes RRAS to resend that packet. Here's how to configure this Registry setting:

1. Launch Regedt32.

2. Select the Window menu option for **HKEY_LOCAL_MACHINE**.

3. Use the tree control in the left-hand window to navigate to the **SYSTEM\CurrentControlSet\Services\Rasman\PPP** subkey. Double-click on the subkey to expand it and display its values in the right-hand window.

4. Locate the **RestartTimer** value entry. Use the DWORD Editor to change the RestartTimer value to a setting that meets the needs of the current network environment.

Setting PPP RRAS Configuration Timeouts

When RRAS uses PPP, it needs guidance on how to configure its networking requests based on the behavior of the local system. These settings are found in the Registry. One Registry setting controls the delay in aborting a connection when protocols cannot be worked out. Here's how to configure this Registry setting:

1. Launch Regedt32.

2. Select the Window menu option for **HKEY_LOCAL_MACHINE**.

3. Use the tree control in the left-hand window to navigate to the **SYSTEM\CurrentControlSet\Services\Rasman\PPP** subkey. Double-click on the subkey to expand it and display its values in the right-hand window.

4. Locate the **NegotiateTime** value entry. Use the DWORD Editor to change the NegotiateTime value to a setting that meets the needs of the current network environment.

Forcing PPP RRAS Password Encryption

Another Registry setting associated with RRAS and PPP controls whether or not to use password encryption when authenticating. Here's how to configure this Registry setting:

1. Launch Regedt32.

2. Select the Window menu option for **HKEY_LOCAL_MACHINE**.

3. Use the tree control in the left-hand window to navigate to the **SYSTEM\CurrentControlSet\Services\Rasman\PPP** subkey. Double-click on the subkey to expand it and display its values in the right-hand window.

4. Locate the **ForceEncryptedPassword** value entry. If the value does not exist, you can create it. Use the DWORD Editor to change the ForceEncryptedPassword value to 1 to force password encryption or 0 to disable it.

Enabling PPP RRAS Logging

Another Registry setting regarding RRAS and PPP controls whether the system should use PPP logging to track problems. Here's how to configure this Registry setting:

1. Launch Regedt32.

2. Select the Window menu option for **HKEY_LOCAL_MACHINE**.

3. Use the tree control in the left-hand window to navigate to the **SYSTEM\CurrentControlSet\Services\Rasman\PPP** subkey. Double-click on the subkey to expand it and display its values in the right-hand window.

4. Locate the **Logging** value entry. If the value does not exist, you can create it. Use the DWORD Editor to change the Logging value to 1 to enable logging and 0 to disable it.

Forcing Clients to Log On Using Dial-Up Networking

If you have network users who work from laptops computers and who frequently use Dial-Up Networking (DUN) to log on to the network, you can configure their systems to dial in automatically by tweaking the Registry. Simply follow these instructions:

1. Launch Regedt32.

2. Select the Window menu option for **HKEY_LOCAL_MACHINE**.

3. Use the tree control in the left-hand window to navigate to the **SOFTWARE\Microsoft\WindowsNT\CurrentVersion\Winlogon** subkey. Double-click on the subkey to expand it and display its values in the right-hand window.

4. Add a new REG_SZ value named **RasForce** and set its value to 1. After the client system reboots, the "Logon Using Dial-Up Networking" checkbox is automatically selected.

Setting the Number of Rings for RRAS to Answer

When remote access clients call your server, you can actually control the number of rings to trigger the server to answer the call. This is controlled with a little known Registry tweak of course. Follow these instructions to make the change:

1. Launch Regedt32.
2. Select the Window menu option for **HKEY_LOCAL_MACHINE**.
3. Use the tree control in the left-hand window to navigate to the **SYSTEM\CurrentControlSet\Services\RasMan\Parameters** subkey. Double-click on the subkey to expand it and display its values in the right-hand window.
4. Add a new REG_DWORD value named **NumberOfRings** and set its value to the number of rings you would like to use. This number can be from 1 to 20.

Keeping a Remote Access Connection up after Logout

When a remote access client logs out, the RRAS server connection with typically terminates. This is often the desired behavior, except in special circumstances in which the remote client needs to offer services to the network after logout. In order to keep these connections alive, tweak the Registry as follows:

1. Launch Regedt32.
2. Select the Window menu option for **HKEY_LOCAL_MACHINE**.
3. Use the tree control in the left-hand window to navigate to the **SOFTWARE\Microsoft\Windows NT\CurrentVersion\Winlogon** subkey. Double-click on the subkey to expand it and display its values in the right-hand window.
4. Add a new REG_SZ value named **KeepRasConnections** and set its value to 1 to keep the connections alive.

Windows 2000 Security

(continued)

In Brief

Unlike other versions of Windows (such as Windows 3.x and 9x), Windows NT/2000 features a complete security system that is an integral part of the OS and cannot be disabled. Windows 2000's security system verifies users' identities when they log onto the system, controls which resources users can access, and maintains an audit trail for security events generated while a user is active. The Registry plays a vital part in all these aspects of Windows 2000 security.

Windows 2000 Security Basics

There are three major aspects of Windows 2000 security: access tokens for users, per-user information granting access rights in the Security Accounts Manager (SAM) Registry keys, and security event logging for potential administrator auditing. These three major aspects of Windows 2000 security, if mastered, can prove to be quite successful for protecting sensitive networks.

Access Tokens

Access tokens are software objects that the Windows 2000 security system creates. They contain information about a user's privileges on both the current workstation and other computers on the network. Access tokens include information that is stored and can be manipulated directly in the Registry. Access tokens can contain the following information:

- User Security ID (SID)
- Group Security ID(s)
- User Privileges
- Owner SID
- Primary Group SID
- Default Access Control List (ACL)

The SAM Registry Keys

Information about users' security settings is stored in the Registry in a subkey of **HKEY_LOCAL_MACHINE** called Security Accounts Manager. SAM is a Windows 2000 service responsible for validating the user logon information. This information typically comes from the user interface of various system utilities. The SAM also provides the information contained in each user's access token. Normally, you do not modify information in the SAM Registry keys directly, but through various administration tools, such as Group Policies.

Auditing

Any security-related action that takes place on a Windows 2000 system has the potential to generate a security event, which in turn can be written into a log file for later auditing by the administrator. You can view this log file using the Event Viewer application, but its settings are in the Registry and you can change them if necessary. Events that can trigger a security event include the following:

- New user creation/group membership changes
- Program execution/resource access
- Logging on/logging off
- Security policy changes
- Privilege usage
- System events affecting security

Logging On

Windows 2000 security comes into play the moment a user attempts to log on to a Windows 2000 computer workstation. The information entered into the Logon dialog box is submitted to the Local Security Authority (LSA), a system service that validates the logon information against the SAM database (either locally or remotely, as discussed in the next section). If the process succeeds, the system creates an access token for the user and keeps it available. This token combines with one or more executable programs to create a *subject* that then begins accessing system capabilities. LSA uses Registry entries to determine how this process works—particularly the user interface, which is shown during the logon process.

The Netlogon Service

Windows 2000 supports a single security database for multiple workstations connected over a network via the Netlogon service. Netlogon requires a domain-based network and the services of a domain controller. Under these conditions, the Logon dialog box displays an extra edit control to permit the entry of a domain name to use for Netlogon verification. There are Registry entries that you can modify directly to control various features of Netlogon.

Shutting Down

Security issues also come into play when it's time for a user to log off a workstation and possibly shut it down. Some users might see only a Logoff dialog box; others see a dialog box with the options of shutting down the workstation or turning off its power. There are Registry entries that control all these features.

Immediate Solutions

Locating the Security Events Log File

You can enable Windows 2000 to write security-related events to a log file. The Registry contains settings that control how the system creates and maintains this log file. To view the log file's location, follow these steps:

1. Launch Regedt32.

2. Select the Window menu option for **HKEY_LOCAL_MACHINE**.

3. Use the tree control in the left-hand window to navigate to the **SYSTEM\CurrentControlSet\Services\EventLog\Security** subkey. Double-click on the subkey to expand it and display its values in the right-hand window.

4. Locate the **File** value entry, which contains the path and file name used to store the security events log file.

Determining the Security Events Log File Maximum Size

MaxSize is another setting that controls the security log file's creation and maintenance. To view the security events log file's maximum size in kilobytes, follow these steps:

1. Launch Regedt32.

2. Select the Window menu option for **HKEY_LOCAL_MACHINE**.

3. Use the tree control in the left-hand window to navigate to the **SYSTEM\CurrentControlSet\Services\EventLog\Security** subkey. Double-click on the subkey to expand it and display its values in the right-hand window.

4. Locate the **MaxSize** value entry. The MaxSize value contains the maximum size in kilobytes the security events log file can reach.

Checking the Lifetime of Security Log Events

You can enable Windows 2000 to write security-related events to a log file. The Registry contains settings that control how the system creates and maintains this log file. To view how long events can remain in the security events log before being overwritten, follow these steps:

1. Launch Regedt32.

2. Select the Window menu option for **HKEY_LOCAL_MACHINE**.

3. Use the tree control in the left-hand window to navigate to the **SYSTEM\CurrentControlSet\Services\EventLog\Security** subkey. Double-click on the subkey to expand it and display its values in the right-hand window.

4. Locate the **Retention** value entry. The Retention value contains the value in seconds that specifies how long events are kept in the log file. Any event "newer" than this value won't be overwritten in the log file, but an event "older" can be overwritten should the log become full. If as a result of checking the Retention and MaxSize settings an event cannot be added to the existing log file, a log full event occurs.

Confirming Whether a Windows 2000 Service Is Logging Security Events

You can determine whether a Windows 2000 service is logging its events into the security log file. As you might guess, you learn this information in the Registry. To check this information, follow these steps:

1. Launch Regedt32.

2. Select the Window menu option for **HKEY_LOCAL_MACHINE**.

3. Use the tree control in the left-hand window to navigate to the **SYSTEM\CurrentControlSet\Services\EventLog\Security** subkey. Double-click on the subkey to expand it and display its values in the right-hand window.

4. Locate the **Sources** value entry. The Sources value contains the names of all services, applications, and groups of applications that are currently logging events into the security events log.

WARNING! The Sources value is dynamic and is maintained by the EventLog service.

Forcing a Windows 2000 System Crash When the Security Events Log Is Full

In an extremely high-security environment, you do not want the Windows 2000 system to keep functioning after the security event log becomes full; otherwise you will miss important security auditing information due to the log full state. Of course, you could always implement this in a far less severe manner using third-party security software for Windows 2000. However, to force a system crash when the log file is full, modify the Registry as follows:

1. Launch Regedt32.

2. Select the Window menu option for **HKEY_LOCAL_MACHINE**.

3. Use the tree control in the left-hand window to navigate to the **SYSTEM\CurrentControlSet\Control\LSA** subkey. Double-click on the subkey to expand it and display its values in the right-hand window.

4. Locate the **CrashOnAuditFail** value entry. Use the DWORD Editor to set the CrashOnAuditFail value to 1 to force a system crash when the security events log cannot be expanded due to a combination of maximum size and event entry lifetime settings.

Locating the Path for **SYSVOL**

SYSVOL is an extremely important shared directory in Windows 2000 that stores the server copy of the domain's public files, which are replicated among all domain controllers in the domain. To access the Registry key that controls the path to **SYSVOL**, follow these steps:

1. Launch Regedt32.

2. Select the Window menu option for **HKEY_LOCAL_MACHINE**.

3. Use the tree control in the left-hand window to navigate to the **SYSTEM\CurrentControlSet\Services\Netlogon\Parameters** subkey. Double-click on the subkey to expand it and display its values in the right-hand window.

4. Locate the **SysVol** value entry. Use the String Editor to check the SysVol value.

Establishing C2-Level Security for CD-ROM Drives

A number of vital Registry entries affect how Windows 2000 behaves at logon. One Registry key determines whether CD-ROM drives and their removable media are accessible only to the currently logged on user (a requirement of Department of Defense C2-level security, the second of four levels of IS system security). To access this key, follow these steps:

1. Launch Regedt32.
2. Select the Window menu option for **HKEY_LOCAL_MACHINE**.
3. Use the tree control in the left-hand window to navigate to the **SOFTWARE\Microsoft\WindowsNT\CurrentVersion\Winlogon** subkey. Double-click on the subkey to expand it and display its values in the right-hand window.
4. Locate the **AllocateCDRoms** value entry. Use the String Editor to change the AllocateCDRoms value to 1. The default for this value is 0 indicating CD-ROM drives can be shared.

Establishing C2-Level Security for Floppy Disk Drives

To access the Registry key that determines whether disk drives and their removable media are accessible only to the currently logged on user, follow these steps:

1. Launch Regedt32.
2. Select the Window menu option for **HKEY_LOCAL_MACHINE**.
3. Use the tree control in the left-hand window to navigate to the **SOFTWARE\Microsoft\WindowsNT\CurrentVersion\Winlogon** subkey. Double-click on the subkey to expand it and display its values in the right-hand window.
4. Locate the **AllocateFloppies** value entry. Use the String Editor to change the AllocateFloppies value to 1.

Establishing C2-Level Security for Floppy Disk Drives and CD-ROMS

If you are serious about C2-level security and want to set it quickly and efficiently for both your floppy drives and your CD-ROM drives, use the following Registry tweak:

1. Launch Regedt32.

2. Select the Window menu option for **HKEY_LOCAL_MACHINE**.

3. Use the tree control in the left-hand window to navigate to the **SOFTWARE\Microsoft\WindowsNT\CurrentVersion\Winlogon** subkey. Double-click on the subkey to expand it and display its values in the right-hand window.

4. Locate the **allocatedasd** value entry. Use the String Editor to change the value to 1. Again, the default is 0 indicating these drives can be shared.

Configuring an Automatic Logon at System Startup

A number of vital Registry entries affect how Windows 2000 behaves at logon. You can configure the Registry to automatically log on a specified user on restart rather than displaying the logon user interface. This allows unattended restarts for remote machines. To do this, follow these steps:

1. Launch Regedt32.

2. Select the Window menu option for **HKEY_LOCAL_MACHINE**.

3. Use the tree control in the left-hand window to navigate to the **SOFTWARE\Microsoft\WindowsNT\CurrentVersion\Winlogon** subkey. Double-click on the subkey to expand it and display its values in the right-hand window.

4. Locate the **AutoAdminLogon** value entry. Use the String Editor to change the AutoAdminLogon value to 1 to authorize automatic logon at startup and 0 to require a manual logon using the user interface.

WARNING! If the AutoAdminLogon value is set to 1, you must also make valid entries in the DefaultUserName and DefaultPassword values, because Windows 2000 uses these values for automatic login at startup. You should also note that these values are not in an area of the Registry that is protected with high security.

Setting the Default User Username

To access the Registry key that determines the username used by the last successfully logged on user (or the username to be used for automatic logon), follow these steps:

1. Launch Regedt32.

2. Select the Window menu option for **HKEY_LOCAL_MACHINE**.

3. Use the tree control in the left-hand window to navigate to the **SOFTWARE\Microsoft\WindowsNT\CurrentVersion\Winlogon** subkey. Double-click on the subkey to expand it and display its values in the right-hand window.

4. Locate the **DefaultUserName** value entry. Use the String Editor to change the DefaultUserName value to specify the user you wish to use for automatic logon at startup. See the later immediate solution that details how to clear this information.

12. Windows 2000 Security

Setting the Default User Password

To access the Registry key that determines the password used by the last successfully logged on user (or the password to be used for automatic logon), follow these steps:

1. Launch Regedt32.

2. Select the Window menu option for **HKEY_LOCAL_MACHINE**.

3. Use the tree control in the left-hand window to navigate to the **SOFTWARE\Microsoft\WindowsNT\CurrentVersion\Winlogon** subkey. Double-click on the subkey to expand it and display its values in the right-hand window.

4. Locate the **DefaultPassword** value entry. To enable automatic logon at startup, use the String Editor to change the DefaultPassword value to the value of the user in **DefaultUserName**.

Determining the Domain of the Last Logged on User

To access the Registry key that determines the domain name of the last user who successfully logged on to the current system, follow these steps:

1. Launch Regedt32.

2. Select the Window menu option for **HKEY_LOCAL_MACHINE**.

3. Use the tree control in the left-hand window to navigate to the **SOFTWARE\Microsoft\WindowsNT\CurrentVersion\Winlogon** subkey. Double-click on the subkey to expand it and display its values in the right-hand window.

4. Locate the **DefaultDomainName** value entry. The DefaultDomainName contains the domain name of the last successfully logged on user.

Deleting Locally Cached Profiles Automatically

To create the Registry key that determines whether or not to delete locally cached user profiles when a roaming user logs off (to conserve disk space and enhance security), follow these steps:

1. Launch Regedt32.

2. Select the Window menu option for **HKEY_LOCAL_MACHINE**.

3. Use the tree control in the left-hand window to navigate to the **SOFTWARE\Microsoft\WindowsNT\CurrentVersion\Winlogon** subkey. Double-click on the subkey to expand it and display its values in the right-hand window.

4. Add the **DeleteRoamingCache** value entry. It is a DWORD value. Set the value to 1 if you want to delete roaming user profiles or to 0 if you want to keep such information on the local hard drive.

Disabling Last Logged on Username Display in the Logon Information Dialog Box

To access the Registry key that determines whether or not to display the username of the last successfully logged on user in the logon user interface, follow these steps:

1. Launch Regedt32.

2. Select the Window menu option for **HKEY_LOCAL_MACHINE**.

3. Use the tree control in the left-hand window to navigate to the **SOFTWARE\Microsoft\WindowsNT\CurrentVersion\Winlogon** subkey. Double-click on the subkey to expand it and display its values in the right-hand window.

4. Locate the **DontDisplayLastUserName** value entry. Use the String Editor to change the DontDisplayLastUserName value to 1 to suppress the display of the last successfully logged on user's username and 0 to restore this feature.

Setting a Windows 2000 Logon Warning Notice Dialog Box Text

There is a Registry key that determines whether the system displays a customized warning dialog box before displaying the Logon dialog box. A user must dismiss the custom dialog box before the system allows him or her to access the Logon dialog box. To access this Registry key, follow these steps:

1. Launch Regedt32.

2. Select the Window menu option for **HKEY_LOCAL_MACHINE**.

3. Use the tree control in the left-hand window to navigate to the **SOFTWARE\Microsoft\WindowsNT\CurrentVersion\Winlogon** subkey. Double-click on the subkey to expand it and display its values in the right-hand window.

4. Locate the **LegalNoticeText** value entry. Use the String Editor to change the LegalNoticeText value to specify the message to be displayed in the custom dialog box.

12. Windows 2000 Security

235

Setting a Windows 2000 Logon Warning Notice Dialog Box Caption

There is a Registry key that determines whether the system displays a customized warning dialog box before displaying the Logon dialog box. A user must dismiss the custom dialog box before the system allows him or her to access the Logon dialog box. To access this Registry key, follow these steps:

1. Launch Regedt32.

2. Select the Window menu option for **HKEY_LOCAL_MACHINE**.

3. Use the tree control in the left-hand window to navigate to the **SOFTWARE\Microsoft\WindowsNT\CurrentVersion\Winlogon** subkey. Double-click on the subkey to expand it and display its values in the right-hand window.

4. Locate the **LegalNoticeCaption** value entry. Use the String Editor to change the LegalNoticeCaption value to the desired caption for the custom dialog box displayed prior to presentation of the Logon dialog box user interface.

Setting a Windows 2000 Logon Information Dialog Box Custom Message

To access the Registry key that determines whether an additional custom message is displayed in the standard Logon dialog box, follow these steps:

1. Launch Regedt32.

2. Select the Window menu option for **HKEY_LOCAL_MACHINE**.

3. Use the tree control in the left-hand window to navigate to the **SOFTWARE\Microsoft\WindowsNT\CurrentVersion\Winlogon** subkey. Double-click on the subkey to expand it and display its values in the right-hand window.

4. Locate the **LogonPrompt** value entry. Use the String Editor to change the LogonPrompt value to the text you want displayed as the prompt in the Logon dialog box.

TIP: If the LogonPrompt feature has never been enabled on the Windows 2000 computer, you will need to create the value using the Edit/Add value.

Setting a Custom Logon Information Dialog Box Additional Caption

To access the Registry key that determines the caption shown in all Logon dialog boxes (in addition to the standard caption), follow these steps:

1. Launch Regedt32.

2. Select the Window menu option for **HKEY_LOCAL_MACHINE**.

3. Use the tree control in the left-hand window to navigate to the **SOFTWARE\Microsoft\WindowsNT\CurrentVersion\Winlogon** subkey. Double-click on the subkey to expand it and display its values in the right-hand window.

4. Add the **Welcome String** value entry. Use the String Editor to add any additional caption text.

Setting a Windows 2000 Logon Password Expiration Warning

You can use the Registry to determine whether a warning dialog box appears when a user logs on and the user's password is within a certain number of days from its expiration date. To do this, follow these steps:

1. Launch Regedt32.

2. Select the Window menu option for **HKEY_LOCAL_MACHINE**.

3. Use the tree control in the left-hand window to navigate to the **SOFTWARE\Microsoft\WindowsNT\CurrentVersion\Winlogon** subkey. Double-click on the subkey to expand it and display its values in the right-hand window.

4. Locate the **PasswordExpiryWarning** value entry. Use the DWORD Editor to change the PasswordExpiryWarning value to the number of days prior to expiration that this warning dialog box is shown after a user logs in with a near-expiration password.

TIP: *The default value for the PasswordExpiryWarning value is 14 days.*

12. Windows 2000 Security

Enabling Shutdown/Power Off Selection in the Shutdown Dialog Box

To access the Registry setting that determines whether an option to shut down the computer and power it off appears in the Shutdown Computer dialog box, follow these steps:

1. Launch Regedt32.

2. Select the Window menu option for **HKEY_LOCAL_MACHINE**.

3. Use the tree control in the left-hand window to navigate to the **SOFTWARE\Microsoft\WindowsNT\CurrentVersion\Winlogon** subkey. Double-click on the subkey to expand it and display its values in the right-hand window.

4. Locate the **PowerdownAfterShutdown** value entry. Use the String Editor to change the PowerdownAfterShutdown value to 1 to enable selecting power off and 0 to disable the feature.

TIP: *The PowerdownAfterShutdown value is 0 in Windows 2000 Server and 1 in Windows 2000 Workstation.*

WARNING! *Not all computers support software-based power shutdowns. Enabling software-based power shutdowns on machines that don't support this feature can result in inconsistent behavior.*

Preventing User Profile Selection Dialog Box Timeouts

To access the Registry key that determines how long a user is allowed to decide between using a local profile and a server profile when there is a delay in connecting to the remote server, follow these steps:

1. Launch Regedt32.

2. Select the Window menu option for **HKEY_LOCAL_MACHINE**.

3. Use the tree control in the left-hand window to navigate to the **SOFTWARE\Microsoft\WindowsNT\CurrentVersion\Winlogon** subkey. Double-click on the subkey to expand it and display its values in the right-hand window.

4. Add the **ProfileDlgTimeOut** value entry. It is a DWORD value. Set it to the number of seconds users are permitted to decide between using a local or a remote server profile if a delay is encountered when contacting a remote server.

TIP: The ProfileDlgTimeOut feature comes into use only if the SlowLinkDetectEnabled value is set to 1.

Enabling Custom Boot Verification

To access the Registry key that determines whether an administrator can install custom bootup verification application(s), follow these steps:

1. Launch Regedt32.
2. Select the Window menu option for **HKEY_LOCAL_MACHINE**.
3. Use the tree control in the left-hand window to navigate to the **SOFTWARE\Microsoft\WindowsNT\CurrentVersion\Winlogon** subkey. Double-click on the subkey to expand it and display its values in the right-hand window.
4. Locate the **ReportBootOK** value entry. Use the String Editor to change the ReportBootOK value to 0 to enable the use of a custom bootup verification application.

TIP: The actual custom bootup verification program is installed via the BootVerificationProgram or BootVerification key.

Enabling Shutdown from the Logon Information Dialog Box

To access the Registry key that determines whether a user can shut down a Windows 2000 computer without actually logging on using the Logon Information dialog box, follow these steps:

1. Launch Regedt32.
2. Select the Window menu option for **HKEY_LOCAL_MACHINE**.

3. Use the tree control in the left-hand window to navigate to the **SOFTWARE\Microsoft\WindowsNT\CurrentVersion\Winlogon** subkey. Double-click on the subkey to expand it and display its values in the right-hand window.

4. Locate the **ShutdownWithoutLogon** value entry. Use the String Editor to change the ShutdownWithoutLogon value to 1 to enable the Shut Down button in the Logon dialog box and 0 to disable it.

Enabling Slow Link Detection

You can use the Registry to determine whether the system looks at the length of time it takes to connect with a remote computer containing user profiles. You can use this information to decide whether to offer users the option of using a local profile rather than a remote one. To do this, follow these steps:

1. Launch Regedt32.

2. Select the Window menu option for **HKEY_LOCAL_MACHINE**.

3. Use the tree control in the left-hand window to navigate to the **SOFTWARE\Microsoft\WindowsNT\CurrentVersion\Winlogon** subkey. Double-click on the subkey to expand it and display its values in the right-hand window.

4. Add the **SlowLinkDetectEnabled** value entry. It is a DWORD value. Set the value to 1 to enable detection of a slow profile link and 0 to disable this feature.

TIP: *The SlowLinkTimeOut and ProfileDlgTimeOut values are required with the SlowLinkDetectEnabled value.*

Preventing Slow Link Timeouts

You can use the Registry to determine the length of time a remote server containing a roaming user's profile can take to connect to the server at which that user is logging on before the system offers the user a local profile. For example, if the link to the remote machine containing the roaming user's profile is down, that user can never log on unless this option is available. To prevent such slow link timeouts, follow these steps:

1. Launch Regedt32.

2. Select the Window menu option for **HKEY_LOCAL_MACHINE**.

3. Use the tree control in the left-hand window to navigate to the **SOFTWARE\Microsoft\Windows2000\CurrentVersion\Winlogon** subkey. Double-click on the subkey to expand it and display its values in the right-hand window.

4. Add the **SlowLinkTimeOut** value entry. It is a DWORD value. Set the SlowLinkTimeOut value to a higher setting to prevent profile linkage timeouts.

TIP: *The SlowLinkTimeOut value can range from 0 to 120,000 milliseconds, with a default of 2,000 (2 seconds). To set this value you must also set the SlowLinkDetectEnabled and ProfileDlgTimeOut values.*

Running an Executable at System Startup

To access the Registry key that determines which executables run in the system context during system initialization, follow these steps:

1. Launch Regedt32.

2. Select the Window menu option for **HKEY_LOCAL_MACHINE**.

3. Use the tree control in the left-hand window to navigate to the **SOFTWARE\Microsoft\WindowsNT\CurrentVersion\Winlogon** subkey. Double-click on the subkey to expand it and display its values in the right-hand window.

4. Locate the **System** value entry. Use the String Editor to add any executables (and paths) you want to run during initialization.

Configuring a Custom Windows 2000 Task Manager

To access the Registry key that determines whether to use a custom Task Manager application, follow these steps:

1. Launch Regedt32.

2. Select the Window menu option for **HKEY_LOCAL_MACHINE**.

3. Use the tree control in the left-hand window to navigate to the **SOFTWARE\Microsoft\WindowsNT\CurrentVersion\Winlogon** subkey. Double-click on the subkey to expand it and display its values in the right-hand window.

4. Add the **Taskman** value entry. It is a String value. You can set the value to the name (and path, if needed) of the custom task manager.

TIP: If there is no entry in the Taskman value or Taskman does not exist (the default), the Windows 2000 Task Manager is used. If the Taskman feature has never been enabled on the Windows 2000 computer, you need to create the value using Edit/Add value.

Running Executables at User Logon

A number of vital Registry entries affect how Windows 2000 behaves at logon. To access the Registry value that determines whether any executable programs run after a user successfully logs on (in the user's context), follow these steps:

1. Launch Regedt32.

2. Select the Window menu option for **HKEY_LOCAL_MACHINE**.

3. Use the tree control in the left-hand window to navigate to the **SOFTWARE\Microsoft\WindowsNT\CurrentVersion\Winlogon** subkey. Double-click on the subkey to expand it and display its values in the right-hand window.

4. Locate the **Userinit** value entry. Use the String Editor to add any executable files (and paths) you want to run after a user logs on.

TIP: The default value of the Userinit value entry contains Userinit (to run the shell) and optionally Nddeagnt.exe (to run NetDDE).

Configuring the Display of the Logoff Dialog Box

When a user prepares to log off in Windows 2000, several Registry entries determine the options shown in the Logoff dialog box. To access the Registry key that specifies these options, follow these steps:

1. Launch Regedt32.

2. Select the Window menu option for **HKEY_CURRENT_USER**.

3. Use the tree control in the left-hand window to navigate to the **SOFTWARE\Microsoft\WindowsNT\CurrentVersion** subkey. Double-click on the subkey to expand it and display its values in the right-hand window.

4. If the **Shutdown** subkey does not exist, add it using the Edit|Add key.

5. Add the **LogoffSetting** value entry. It is a DWORD value. Set the value to one of those in Table 12.1.

Table 12.1   Setting the Logoff dialog box behavior.

Value	Logoff Dialog Box Behavior
0	Log off dialog box
1	Shut down dialog box
2	Shut down and restart
3	Shut down and power off (when supported)

Configuring the Display of the Shutdown Dialog Box

When a user prepares to log off in Windows 2000, several Registry entries determine the options shown in the Shutdown dialog box. To access the Registry key that specifies these options, follow these steps:

1. Launch Regedt32.

2. Select the Window menu option for **HKEY_CURRENT_USER**.

3. Use the tree control in the left-hand window to navigate to the **SOFTWARE\Microsoft\WindowsNT\CurrentVersion** subkey. Double-click on the subkey to expand it and display its values in the right-hand window.

12. Windows 2000 Security

243

4. If the **Shutdown** subkey does not exist, add it using the Edit|Add key.

5. Add the **ShutdownSetting** value. It is a DWORD value. Set the value to one of those shown in Table 12.2.

Table 12.2   Setting the Shut Down dialog box behavior.

Value	Shut Down Dialog Box Behavior
0	Log off dialog box
1	Shut down dialog box
2	Shut down and restart
3	Shut down and power off (when supported)

Clearing the Pagefile at Shutdown

It is a little known fact that the Windows 2000 system's pagefile is left intact when the system shuts down, a situation that a computer criminal can exploit. To eliminate this possibility, follow these steps:

1. Launch Regedt32.

2. Select the Window menu option for **HKEY_LOCAL_MACHINE**.

3. Use the tree control in the left-hand window to navigate to the **SYSTEM\CurrentControlSet\Control\Session Manager\ Memory Management** subkey. Double-click on the subkey to expand it and display its values in the right-hand window.

4. Locate the **ClearPageFileAtShutdown** value and set it to 1.

WARNING! This increases the time it takes your Windows 2000 system to shut down in direct proportion to the size of the pagefile.

Preventing the Caching of Logon Credentials

Windows 2000 Professional caches the last 10 sets of logon credentials received from a domain controller. This can increase network efficiency, but might not be desirable in a high-security environment. Here are the steps for preventing this behavior:

1. Launch Regedt32.

2. Select the Window menu option for **HKEY_LOCAL_MACHINE**.

3. Use the tree control in the left-hand window to navigate to the **SOFTWARE\Microsoft\Windows NT\CurrentVersion\Winlogon** subkey. Double-click on the subkey to expand it and display its values in the right-hand window.

4. Add a REG_SZ value named **CachedLogonsCount** and set its value to 0.

Chapter 13

Windows 2000 Help

In Brief

Windows 2000 is the first major operating system release from Microsoft to depend entirely on the new Hypertext Help system, rather than the older WinHelp format. Because Hypertext Help is HTML-based, it can include active content that Windows 2000 leverages into its powerful troubleshooter system. At the same time, Hypertext Help keeps the highly useful context-sensitive help features and What's This? from the WinHelp era. A number of key aspects of Windows 2000 Help depend on Registry entries, and are outlined in this chapter's "Immediate Solutions" section.

Windows 2000 Help Features

Here, we'll discuss many of Windows 2000 Help Features.

Troubleshooters

The Windows 2000 Hypertext Help system includes a powerful new feature called *troubleshooters*. You may access troubleshooters via an application's Hypertext Help, or from the main Windows 2000 Help system to help you with a number of topics, including the following:

- 16-bit Windows
- Blue screens (crashes)
- Hardware compatibility
- Networks
- Modems
- Displays
- TCP/IP

Figure 13.1 shows a troubleshooter in operation.

What's This?

Another Help feature brought over from earlier Windows Help systems is the What's This? feature, which usually manifests as a small button near the Close, Minimize, and Maximize buttons on dialog boxes. Clicking on the What's This? button opens a special cursor, and when you click on a user interface element, a pop-up Help window opens and gives a succinct description of the particular element. Figure 13.2 shows What's This? in operation.

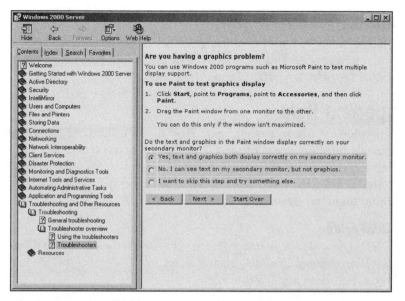

Figure 13.1 Hypertext Help troubleshooters in Windows 2000.

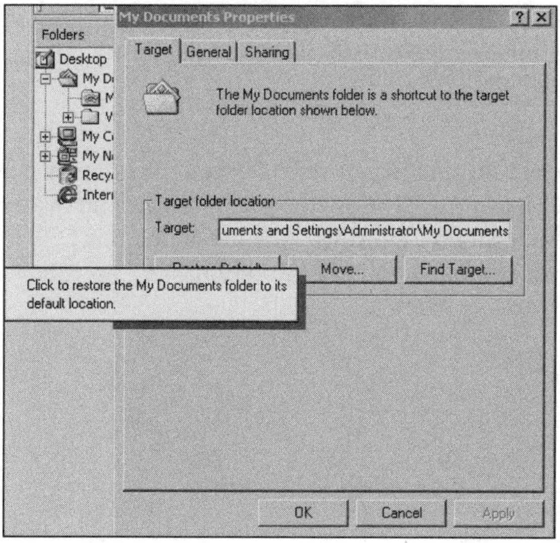

Figure 13.2 What's This? in Windows 2000.

Context-Sensitive Help

One of the nicer features of the older WinHelp system is its ability to jump to a precise spot in a large Help file, depending on where you asked for help. Hypertext Help retains this ability.

Hypertext Help

The older WinHelp system used HLP files exclusively. Most users quickly became aware of the file extension and learned to avoid removing or moving HLP files. Hypertext Help uses three different files types: CHM, CHI, and DSZ. Each file type serves a different role in the Hypertext Help system.

CHM Files

CHM (Compiled Hypertext Material) files are the main data files for the Help system. Double-clicking on a CHM file opens the file in the Hypertext Help application.

CHI Files

CHI files are the index files for the CHM documents. CHI files produce the left-hand tree control in the Hypertext Help viewer. If you delete or move a CHI file, its main CHM file will still open, but it won't have any external navigation capabilities.

DSZ Files

DSZ files are the script files for Windows 2000 troubleshooters. They interact with the CHM and CHI files depending on the answers to questions in the troubleshooter interface. Moving or deleting DSZ files cripples the ability of troubleshooters to function.

Immediate Solutions

Locating the Troubleshooters Installation Directory

Almost all the troubleshooters for Windows 2000 live in one spot. If this one folder gets renamed, moved, or deleted, the troubleshooters will promptly stop working. To restore lost troubleshooters for any of these reasons, administrators can tweak the Registry as follows:

1. Launch Regedt32.

2. Select the Window menu option for **HKEY_LOCAL_MACHINE**.

3. Use the tree control in the left-hand window to navigate to the **Software\Microsoft\Tshoot** subkey. Double-click on the subkey to expand it and display its values in the right-hand window.

4. Examine the **FullPathToResource** value entry. Make sure that FullPathToResource points to a valid location. If not, either change the location, or restore the contents of the folder from backup.

Fixing a Broken Troubleshooter Linkage

Troubleshooters are actually a group of files working together, bound by a Registry entry. Removing any of the files or changing the Registry entry breaks this connection. Fortunately, administrators can easily restore the linkage using the following technique:

1. Launch Regedt32.

2. Select the Window menu option for **HKEY_LOCAL_MACHINE**.

3. Use the tree control in the left-hand window to navigate to the **SOFTWARE\Microsoft \Tshoot\TroubleshooterList*[name]*** subkey, where *[name]* is the name of the troubleshooter, which is malfunctioning. Double-click on the subkey to expand it and display its values in the right-hand window.

4. Examine the **Fname** and **Path** value entries. Ensure that they point to existing locations and files. If not, restore the files from backup and/or change the path entry appropriately.

Verifying a Third-Party Troubleshooter Installation

Microsoft has made the format of its troubleshooter system available to third-party developers of Hypertext Help, making it likely that third-party products will include the capability. If a third-party trouble-shooter fails to function properly, administrators can start the solution process by determining whether the troubleshooter installed correctly:

1. Launch Regedt32.
2. Select the Window menu option for **HKEY_LOCAL_MACHINE**.
3. Use the tree control in the left-hand window to navigate to the **Software\Microsoft\Tshoot\TroubleShooterList** subkey. Double-click on the subkey to expand it and display its values.
4. Check the **TroubleShooterList** subkeys for the name of the third-party product's troubleshooter. If you don't find it, the troubleshooter was never installed or the installation failed. Otherwise, use the procedures described in the previous Immediate Solution to locate the difficulty.

Determining if the Local Troubleshooter Facility Is Installed

Some of the Hypertext Help files for Windows 2000 support a *local troubleshooter*—that is, one that can be invoked directly, rather than from the main troubleshooter list in the principal help file. This allows you to customize the Help system for your particular network. What isn't widely known, is that this capability depends on a Registry entry to load the COM server to provide this feature. If users run into trouble with local troubleshooters, here's how to check the Registry:

1. Launch Regedt32.
2. Select the Window menu option for **HKEY_CLASSES_ROOT**.

3. Use the tree control in the left-hand window to navigate to the **TSHOOT.TSHOOTCtrl.1\CLSID** subkey. Double-click on the subkey to expand it and display its values in the right-hand window.

4. Copy the value of the CLSID key and search for it in the HKCR hive. After you find it, check its **InProcServer32** subkey for a valid path and determine whether the DLL is indeed where it should be. If not, change the path appropriately.

Chapter 14

System Multimedia

In Brief

Multimedia (video, sound, 3D rendering, and so forth) was at one time the province of application vendors and graphics card makers. Then, in the mid-1990s, Microsoft decided to make Windows a powerful multimedia platform and created the technology that has become DirectX. By now the Windows operating system supports almost all types of multimedia, including Windows 2000's Media Player, Audio services, Speech API, and DirectX systems, all of which have critical Registry entries that administrators should know about.

Media Player

Media Player is the latest incarnation of Microsoft's video playback system (also known as ActiveMovie, DirectShow, and several other names). It has a number of powerful features, including the following:

- Automatic codec (video software decoder) download
- Support for AVI, MPEG, and MOV file formats
- Scalable display ranging from compact to full screen
- Stereo sound support
- VCR-style display controls
- Free upgrades

You can configure several essential Media Player elements directly via the Registry. Figure 14.1 shows Media Player in operation.

Figure 14.1   Media Player in Windows 2000.

Audio Services

Along with video comes sound and Windows 2000 features default drivers to play WAV and MPEG audio files, which you can configure from the Registry. There are also settings to interact with the older, non-DirectX sound drivers that can be important for legacy applications.

Speech

Speech, both computer-synthesized and as input from a microphone, is a relative newcomer to the multimedia domain. Windows 2000 keeps critical information about its Speech API in Registry entries. Microsoft's showcase speech demo is the amusing but quite powerful Agent control; you can program it to accept spoken commands, and to interface with a database to provide quite sophisticated user interface behavior.

DirectX

DirectX is Microsoft's name for its multimedia API; it evolved from the older MCI (Media Control Interface) and several similar technologies. The four major types of DirectX technology are the following:

- *DirectPlay*—A facilitating technology for playing games over a network, either the Internet or a LAN. DirectPlay keeps track of available game information via the Registry.

- *DirectMusic*—The technology that creates synthesized music directly (synthesizers used to be called MIDI) rather than simply playing a recording. DirectMusic has all the information about which synthesizers it supports in Registry values.

- *Direct3D*—As its name implies, Direct3D is the technology for creating illusory three-dimensional objects on the screen directly from mathematical information. Direct3D needs a lot of hardware support to function, and it keeps this information in Registry entries.

- *DirectDraw*—DirectDraw takes care of speeding up simple graphics drawing operations, especially bitmap processes. It doesn't keep much information in the Registry, mainly because it is now mostly kept in graphics cards directly, rather than in software.

Immediate Solutions

Checking for Direct3D Hardware Support

In some cases, it may be essential to determine if a Windows 2000 computer has hardware support for Direct3D (such as for visual modeling tasks). An administrator can find out by checking the Registry as follows:

1. Launch Regedt32.

2. Select the Window menu option for **HKEY_LOCAL_MACHINE**.

3. Use the tree control in the left-hand window to navigate to the **SOFTWARE\Microsoft\Direct3D\Drivers\Direct3D HAL** subkey. Double-click on the subkey to expand it and display its values in the right-hand window.

4. If you see any values for the HAL key, the video card supports Direct3D. If not, the video card does not support Direct3D.

Determining if an Application Uses DirectDraw

In some cases, it may be essential to determine if a Windows 2000 application has support for DirectDraw (such as for visual modeling tasks). An administrator can check the Registry to find out. First, run the application in question, then, follow these steps:

1. Launch Regedt32.

2. Select the Window menu option for **HKEY_LOCAL_MACHINE**.

3. Use the tree control in the left-hand window to navigate to the **SOFTWARE\Microsoft\DirectDraw\MostRecentApplication** subkey. Double-click on the subkey to expand it and display its values in the right-hand window.

4. If you see the application name as the **Name** value for this key, then the application supports DirectDraw. If not, the application does not support DirectDraw.

Fixing Broken DirectMusic Synthesizer Links

DirectMusic is a powerful synthesizer capability supported by Windows 2000, but it has an Achilles' heel—it needs a Registry-based path to its synthesizer list. If this path gets changed without updating the Registry, DirectMusic will break. Here's how an administrator can fix the problem:

1. Launch Regedt32.

2. Select the Window menu option for **HKEY_LOCAL_MACHINE**.

3. Use the tree control in the left-hand window to navigate to the **SOFTWARE\Microsoft\DirectMusic** subkey. Double-click on the subkey to expand it and display its values in the right-hand window.

4. Locate the **GMFilePath** value entry. Check to ensure it is correct, and, if not, either move the file to the named directory path, or use the String Editor to change the path to the new one.

Finding Out if a DirectPlay Connection Type Is Supported

More and more, workstation-level computers are being used to test powerful network-oriented gaming software. Windows 2000 DirectPlay supports this process natively, but users might set up their gaming software and attempt a connection only to find the system ignoring them. Administrators can check the Registry to make sure that the needed connection type for a given DirectPlay session is supported by following these steps:

1. Launch Regedt32.

2. Select the Window menu option for **HKEY_LOCAL_MACHINE**.

3. Use the tree control in the left-hand window to navigate to the **SOFTWARE\Microsoft\DirectPlay\ServiceProviders** subkey. Double-click on the subkey to expand it and display its values in the right-hand window.

4. There are as many subkeys to the ServiceProviders key as there are supported connection types. If the type of connection you need is not shown as a key, you need to rerun the DirectPlay installation utility and set up the necessary key.

Setting Media Player's Codec Upgrade Download URL

Media Player, the video system utility for Windows 2000, has a powerful feature that enables it to attempt to download a new codec module when a video file needs one. The "gotcha" is that the URL for the Microsoft codec Web page continually changes. Savvy administrators can keep in touch with the latest URL and update their copies of Media Player by tweaking the Registry as follows:

1. Launch Regedt32.

2. Select the Window menu option for **HKEY_LOCAL_MACHINE**.

3. Use the tree control in the left-hand window to navigate to the **SOFTWARE\Microsoft\MediaPlayer\PlayerUpgrade** subkey. Double-click on the subkey to expand it and display its values in the right-hand window.

4. Locate the **UpgradeServer** value entry. Use the String Editor to change the UpgradeServer value to the new URL for codec download.

Locating the Speech API Installation Directory

Windows 2000 now comes with support for the potent Microsoft speech synthesis and recognition technology called Speech API. Unfortunately, this very useful capability is not well documented, which can result in its DLLs getting moved or deleted accidentally. In order to know where to reinstall them if necessary, check the Registry as follows:

1. Launch Regedt32.

2. Select the Window menu option for **HKEY_LOCAL_MACHINE**.

3. Use the tree control in the left-hand window to navigate to the **SOFTWARE\Microsoft\SpeechApi** subkey. Double-click on the subkey to expand it and display its values in the right-hand window.

4. Locate the **InstallDir** value entry and ensure the DLLs and other necessary files are restored there or are still present.

14. System Multimedia

Finding the Currently Installed Media Control Interface (MCI) Drivers for a Multimedia Type

Before there was DirectX, there was the MCI (Media Control Interface). A lot of multimedia problems can crop up because an MCI driver for a given file type is incorrect. Learning which drivers are in use, requires a clever peek at the Registry as described here:

1. Launch Regedt32.

2. Select the Window menu option for **HKEY_LOCAL_MACHINE**.

3. Use the tree control in the left-hand window to navigate to the **SOFTWARE\Microsoft\Windows NT\CurrentVersion\MCI** subkey. Double-click on the subkey to expand it and display its values in the right-hand window.

4. Locate the value entry for the file type desired; it will give the **DRV** file associated with it. Use the String Editor to change the value to the appropriate one if needed.

TIP: Notice that the MCI values are not file extensions; they are MCI device type names.

Locating the Default Windows WAV Audio Driver

Windows has played WAV audio since version 3.1, and sometimes that capability is still important, particulary for legacy applications. To find out what old-style WAV file driver is in use, check the Registry as follows:

1. Launch Regedt32.

2. Select the Window menu option for **HKEY_LOCAL_MACHINE**.

3. Use the tree control in the left-hand window to navigate to the **SOFTWARE\Microsoft\Windows NT\CurrentVersion\ Userinstallable.drivers** subkey. Double-click on the subkey to expand it and display its values in the right-hand window.

4. Locate the **wave** value entry. If the driver file given is incorrect, use the String Editor to change it to the correct one.

14. System Multimedia

Chapter 15

Registry Programming

In Brief

The minute you really start to get serious about Windows 2000 programming is more than likely the same minute you need to learn about programming for the Windows 2000 Registry. Fortunately, Microsoft makes this pretty straightforward thanks to tools that come with the operating system. This chapter is not a complete guide to the vast topic of Windows 2000 Registry programming, but it is more than enough to get you started in the right direction—and for this topic that makes a big difference.

This chapter provides information about the three main methods you can use to enter the world of Registry programming. Note that this chapter assumes you know how to program in C/C++ or Visual Basic. Information on how to do this is obviously beyond the scope of this book. Let's begin by looking at these three methods for programming against the Windows 2000 Registry.

The Windows 2000 Registry API

Microsoft provides a powerful application programming interface to assist developers in accessing the power of the Windows 2000 Registry. This API resides in the Microsoft Developer Network (MSDN) Platform SDK (Software Development Kit). Routines that a programmer can call to manipulate the Registry are the crux of the API. Table 15.1 details these routines for you.

Table 15.1   Registry API routines.

Routine	Description
RegCloseKey	Closes a connection between the application and the Registry
RegConnectRegistry	Creates a Registry connection with a remote computer
RegCreateKey	Creates a new Registry subkey
RegCreateKeyEx	Creates a new Registry subkey and permits options to be set
RegDeleteKey	Deletes an existing subkey

(continued)

Table 15.1 Registry API routines *(continued)*.

Routine	Description
RegDeleteValue	Deletes an existing value
RegEnumKey	Enumerates all subkeys specified—compatible with previous Windows versions
RegEnumKeyEx	Enumerates all subkeys specified—used for Windows 2000
RegEnumValue	Enumerates all values in the specified subkey
RegFlushKey	Causes changes to be written to the Registry
RegGetKeySecurity	Obtains the security attributes for a Registry object
RegLoadKey	Creates a new subkey based on information in a file
RegNotifyChangeKeyValue	Causes the system to notify the application if a Registry object is changed
RegOpenKey	Opens a Registry object
RegOpenKeyEx	Opens a Registry object—used for Windows 2000
RegQueryInfoKey	Returns information about a Registry object
RegQueryMultipleValues	Returns information about Registry values in a particular subkey
RegQueryValue	Returns the value of the default value entry for a subkey
RegQueryValueEx	Returns the value of the default value entry for a subkey—used for Windows 2000
RegReplaceKey	Causes the OS to use a different file for a key
RegRestoreKey	Restores the subkey's contents to a file
RegSaveKey	Saves the subkey to a file
RegSetKeySecurity	Sets a Registry object's security attributes
RegSetValue	Sets the value of the default value entry for a key
RegSetValueEx	Sets the value of the default value entry for a key—used for Windows 2000
RegUnLoadKey	Removes the specified object from the Registry
RegOpenUserClassesRoot	Retrieves the **HKEY_CLASSES_ROOT** hive for a specified user account
RegOpenCurrentUser	Retrieves a handle to **HKEY_CURRENT_USER**
RegOverridePredefKey	Allows mapping one key to a different key

15. Registry Programming

These Registry API routines all return error codes as their values. Understand that many of these codes are actually reporting successes. This makes it very easy to test for the success or failure of operations in your application. Table 15.2 lists the most common codes.

Table 15.2   Registry API error codes.

Code	Meaning
ERROR_SUCCESS	Operation success
ERROR_FILE_NOT_FOUND	Key or path does not exist
ERROR_ACCESS_DENIED	Permissions do not allow access to the key
ERROR_INVALID_HANDLE	Key passed is not valid
ERROR_OUTOFMEMORY	Not enough memory
ERROR_INVALID_PARAMETER	Invalid parameters have been passed
ERROR_BAD_PATHNAME	Path does not exist
ERROR_LOCK_FAILED	Internal locking mechanism in the Registry failed
ERROR_MORE_DATA	Buffer provided is too small
ERROR_NO_MORE_ITEMS	No more keys or values to enumerate
ERROR_REGISTRY_RECOVERED	One or more hives was reconstructed
ERROR_KEY_DELETED	Key that is being modified has been deleted
ERROR_CANTREAD	Key can be opened but not read
ERROR_CANTOPEN	Key cannot be opened

Shell Lightweight API

The *Shell Lightweight API* is a more recent API that permits Registry modifications. The Shell Lightweight library is available with Windows 98, Windows 95, Windows NT 4, and Internet Explorer 5 or higher. Shell Lightweight API features new functions that internally open and close specified keys when asked to read or write values. Simply call **SHGetValue** or **SHSetValue** and the functions interact appropriately with the Registry. The new Shell Lightweight API also provides a **SHDeleteKey** function that recursively deletes nonempty keys.

Visual Basic Object Models

Visual Basic is most likely the simplest way in which to program for the Windows 2000 Registry, thanks to a powerful object library called

RegObj. You can download this library at: **http://msdn.microsoft.com/vbasic/downloads**. RegObj is implemented as RegObj.dll— RegObj.dll is a powerful ActiveX server that allows Visual Basic developers to programmatically control the Registry without having to resort to the Windows APIs mentioned above.

15. Registry Programming

Immediate Solutions

Opening a Key Using the Registry API

The recommended method for opening a Registry key in Windows 2000 is to use **RegOpenKeyEx**. The syntax for this routine is as follows:

```
LONG RegOpenKeyEx(hKey, pszSubKey, dwOptions, samDesired,
phkResult);
```

Where:

- **hKey**—Handle to open key or root key

- **pszSubKey**—Name of the subkey of **hKey** you want to open

- **dwOptions**—Reserved; this must be 0

- **samDesired**—Mask that defines the access rights (**KEY_READ** or **KEY_WRITE**)

- **phkResult**—Pointer to the newly opened key

Here is an example as used in Visual Basic. This example demonstrates the setting of a Registry value after the key is opened:

```
Private Sub SetKeyValue (sKeyName As String, sValueName As _
String, vValueSetting As Variant, lValueType As Long)
     Dim lRetVal As Long  'result of the SetValueEx function
     Dim hKey As Long         'handle of open key

     'open the specified key
     lRetVal = RegOpenKeyEx(HKEY_CURRENT_USER, sKeyName, 0, _
                               KEY_SET_VALUE, hKey)
     lRetVal = SetValueEx(hKey, sValueName, lValueType, _
     vValueSetting)
     RegCloseKey (hKey)
     End Sub
```

A call of

```
SetKeyValue "TestKey\SubKey1", "StringValue", "Hello", REG_SZ
```

creates a value of type **REG_SZ** called **StringValue** with the setting of **Hello**. This value will be associated with the key **SubKey1** of **TestKey**.

Creating a Key Using the Registry API

Creating a key is also straightforward thanks to the API. Be sure you have permissions to the area where you would like to create the key, and then usd-the following syntax:

```
LONG RegCreateKeyEx (hKey, pszSubKey, Reserved, pszClass,
dwOptions, samDesired, lpSecurityAttributes, phkResult,
lpdwDisposition);
```

- **hKey**—Handle to open key or root key

- **pszSubKey**—Name of the subkey of **hKey** you want to open

- **Reserved**—Must be NULL

- **pszClass**—Specifies the class of the key

- **dwOptions**—REG_OPTION_NON_VOLATILE or REG_OPTION_VOLATILE or REG_OPTION_BACKUP_RESTORE

- **samDesired**—Mask that defines the access rights (**KEY_READ** or **KEY_WRITE**)

- **lpSecurityAttributes**—Points to a **SECURITY_ATTRIBUTES** structure

- **phkResult**—Pointer to **hKey** containing the newly opened key

- **lpdwDisposition**—Points to a DWORD that indicates what happened

Here is an example of this call using Visual Basic programming:

```
Private Sub CreateNewKey (sNewKeyName As String, lPredefinedKey _
As Long)
Dim hNewKey As Long
Dim lRetVal As Long
lRetVal = RegCreateKeyEx(lPredefinedKey, sNewKeyName, 0&, _
                vbNullString, REG_OPTION_NON_VOLATILE, _
                KEY_ALL_ACCESS, _
                0&, hNewKey, lRetVal)
RegCloseKey (hNewKey)
End Sub
```

Therefore, a procedure call of

```
CreateNewKey "TestKey", HKEY_LOCAL_MACHINE
```

creates a key called **TestKey** immediately under **HKEY_LOCAL_MACHINE**.

Retrieving Information about Keys Using the Registry API

You can obtain an incredible amount of information about a Registry key using the **RegQueryInfoKey** function. This function returns 11 data components for a particular key. Here is the syntax for the function:

```
LONG RegQueryInfoKey (hKey, lpClass, lpcbClass, lpReserved,
lpcSubKeys, lpcbMaxSubKeyLen, lpcbMaxClassLen, lpcValues,
lpcbMaxValueNameLen, lpcbMaxValueLen, lpcbSecurityDescriptor,
lpftLastWriteTime);
```

In the above example, here is what the various components do:

- **hKey**—Handle to open key or root key
- **lpClass**—Points to a buffer that receives the key's class name
- **lpcbClass**—Contains the length of the class name passed back in **lpClass**
- **lpReserved**—This value is reserved and, as such, should always be NULL
- **lpcSubKeys**—Reports the number of subkeys
- **lpcbMaxSubKeyLen**—Reports the length of the longest subkey name
- **lpcbMaxClassLen**—Reports the length of the longest class name
- **lpcValues**—Reports the number of values
- **lpcbMaxValueNameLen**—Reports the length of the longest value name
- **lpcbMaxValueLen**—Reports the length of the longest value contents
- **lpcbSecurityDescriptor**—Reports the size of the security descriptor associated with the key
- **lpftLastWriteTime**—The date and time **hKey** or any of its values were modified

Retrieving a Registry Value Using the Registry API

The **RegQueryValueEx** function is a simple way to retrieve a single value from the Registry. The syntax for this command follows:

```
LONG RegQueryValueEx (hKey, pszValueName, lpReserved, lpType,
lpData, lpcbData);
```

In the above example, here is what the various components do:

- **hKey**—Handle to open key or root key

- **pszValueName**—Name of the value you want to query

- **lpReserved**—This value is reserved, and as such, should always be NULL

- **lpType**—Reports the data type of the value

- **lpData**—Reports the value's contents

- **lpcbData**—Holds amount of data copied into **lpData**

15. Registry Programming

Chapter 16

COM+

In Brief

COM+ is a powerful new version of Microsoft's veteran *Component Object Model (COM)* technology. COM+ puts a number of previously standalone technologies such as Microsoft Transaction Server (MTS) and Microsoft Messaging Queue (MSMQ) directly into the operating system, starting with Windows 2000. Although it has a very nice user interface (called the *Component Services Explorer*) for administration, COM+ also depends on a set of Registry entries, which administrators can use to get the most out of a given server's COM+ installation. Interestingly enough, unlike standard COM, COM+ does not keep its data in the standard Registry. Instead, COM+ keeps its data in a proprietary database called the RegDB. COM+ provides Windows 2000 developers with a wide range of services, including Distributed Transactions, DCOM (Distributed COM) support, Asynchronous Procedure Calls, In-Memory Databases, object pooling, and load balancing.

Power under the Hood

COM+ provides a group of potent services to the Windows 2000 system, including the following:

- Transactions
- Load balancing and failover support
- Object pooling and JIT (Just In Time) object activation
- Asynchronous function calls
- Publish and subscribe events

Figure 16.1 shows the COM+ system in operation.

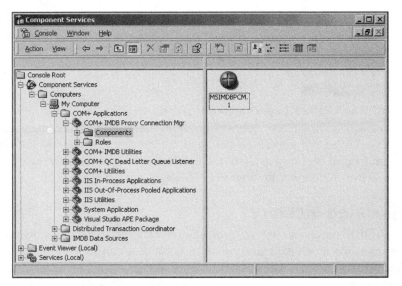

Figure 16.1   COM+ in Windows 2000.

DCOM Support

COM+ provides full support for remote component interactions using Distributed COM (DCOM). You can use a special Registry tweak to control this manually, if necessary. In addition, a new protocol for DCOM called *Component Internet Services (CIS)* allows COM+ components to connect over standard Internet linkages. DCOM itself continues to be administered via DCOMCFG rather than Component Services.

ODBC Resource Pooling

One of the more memory-intensive and time-costly software objects to keep around is an *Open Database Connectivity (ODBC)* database connection—and busy servers can have many of them coming and going at any given time. COM+ enhances performance substantially by keeping ODBC connection resources in memory, even when they aren't being used, provided a Registry value is set correctly.

IMDB

In-Memory Databases (IMDB) is one of COM+'s most useful capabilities. IMDB allows data from a server to be placed in memory and then treated like a local ODBC datasource, complete with the ability to be opened, have SQL queries run against it, and so on. There are three principal features of IMDB that Registry entries control—connections, allocated memory, and BLOBs.

Connections

IMDB works just like a standard ODBC-compliant database server, so it has a concurrent connection limit. Any additional requests received after the maximum connections are reached will fail until a connection becomes available. Administrators can manually raise the concurrent connection limit via a Registry setting.

Allocated Memory

All IMDB databases use a certain amount of hardwired memory, which limits how much data any one of the databases can have. Administrators can raise or lower this limit via the Registry to achieve maximum performance.

BLOBs

Binary Large Objects (BLOBs) are special database fields that hold binary data (such as images). IMDB has a rather lower default size limit for BLOB data. Administrators can increase the size limit if needed for a given database environment by setting a Registry value.

Immediate Solutions

Enabling COM+ Registry Database Automatic Backup

COM+ depends on the Registry for some critical control settings, including whether to automatically back up its proprietary database. To force this backup operation, set the Registry as follows:

1. Launch Regedt32.

2. Select the Window menu option for **HKEY_LOCAL_MACHINE**.

3. Use the tree control in the left-hand window to navigate to the **SOFTWARE\Microsoft\COM3** subkey. Click on the subkey to select it and display its values in the right-hand window.

4. Locate the **RegDBAutoBackup** value entry. Use the DWORD Editor to change the RegDBAutoBackup value to 1 to force automatic backups.

Adding a Remote Computer to the COM+ System Manually

The Registry also includes the names and configuration information of all computers administered by a given COM+ installation. Although you can administer these systems with the CS Explorer, in some cases it might be easier to manually add a computer to the current COM+ system as described in the following steps:

1. Launch Regedt32.

2. Select the Window menu option for **HKEY_LOCAL_MACHINE**.

3. Use the tree control in the left-hand window to navigate to the **SOFTWARE\Microsoft\COM3\ComputerListTable** subkey. Click on the subkey to select it and display its values in the right-hand window.

4. Add a new subkey with the name of the computer you want to add. Then, configure an entry for the new subkey under the **LocalComputerTable** subkey, duplicating one of the subkeys already there.

16. COM+

Enabling DCOM Support in COM+

You can also control whether DCOM support is enabled. If DCOM starts failing for COM+, follow these steps to check the Registry and reset it if necessary:

1. Launch Regedt32.

2. Select the Window menu option for **HKEY_LOCAL_MACHINE**.

3. Use the tree control in the left-hand window to navigate to the **SOFTWARE\Microsoft\COM3\LocalComputerTable\ MyComputer** subkey. Click on the subkey to select it and display its values in the right-hand window.

4. Locate the **EnableDCOM** value entry. Use the String Editor to change the EnableDCOM value to Y to enable DCOM support.

TIP: *You still must use the DCOMCFG utility to configure access for the remote component.*

Enabling COM+ Security Tracking

You can cause COM+ to log all its security activities. You can turn this on using the following Registry tweak:

1. Launch Regedt32.

2. Select the Window menu option for **HKEY_LOCAL_MACHINE**.

3. Use the tree control in the left-hand window to navigate to the **SOFTWARE\Microsoft\COM3\LocalComputerTable\ MyComputer** subkey. Click on the subkey to select it and display its values in the right-hand window.

4. Locate the **EnableSecurityTracking** value entry. Use the String Editor to change the EnableSecurityTracking value to Y to turn on COM+ security event logging.

WARNING! *COM+ uses a special security system called Roles. Be sure you understand Roles before you start using COM+ security features.*

Preventing IMDB Connection Failures

You can control how many concurrent IMDB connections are allowed. If a heavy-traffic situation develops, the administrator can increase this value by changing the Registry as follows:

1. Launch Regedt32.

2. Select the Window menu option for **HKEY_LOCAL_MACHINE**.

3. Use the tree control in the left-hand window to navigate to the **SOFTWARE\Microsoft\COM3\LocalComputerTable\ MyComputer** subkey. Click on the subkey to select it and display its values in the right-hand window.

4. Locate the **IMDBConnections** value entry. Use the DWORD Editor to change the IMDBConnections value to the number of connections desired, remembering that each connection consumes the fixed IMDB memory allowance regardless of how much it needs.

TIP: The default value of IMDBConnections is 10.

Improving IMDB Performance with Dynamic Table Loading

You also have the ability to dynamically load tables while providing results to users. You can turn on this very useful feature by making the following Registry change:

1. Launch Regedt32.

2. Select the Window menu option for **HKEY_LOCAL_MACHINE**.

3. Use the tree control in the left-hand window to navigate to the **SOFTWARE\Microsoft\COM3\LocalComputerTable\ MyComputer** subkey. Click on the subkey to select it and display its values in the right-hand window.

4. Locate the **IMDBLoadTablesDynamically** value entry. Use the String Editor to change its value to Y to turn on this feature.

16. COM+

Improving IMDB Performance by Increasing Available Memory

You can also configure the total memory allocated to IMDB operations on a per-connection basis. In a heavy-traffic situation, administrators can increase this value by changing the Registry as follows:

1. Launch Regedt32.

2. Select the Window menu option for **HKEY_LOCAL_MACHINE**.

3. Use the tree control in the left-hand window to navigate to the **SOFTWARE\Microsoft\COM3\LocalComputerTable\ MyComputer** subkey. Click on the subkey to select it and display its values in the right-hand window.

4. Locate the **IMDBMemorySize** value entry. Use the DWORD Editor to change the IMDBMemorySize value to the desired size (in MB) for each IMDB connection.

TIP: The default value of IMDBMemorySize is 512.

Allowing Large IMDB BLOBs

You can configure how much memory a BLOB can occupy without generating an error. Administrators can change this value depending on specific database environments by altering the Registry as follows:

1. Launch Regedt32.

2. Select the Window menu option for **HKEY_LOCAL_MACHINE**.

3. Use the tree control in the left-hand window to navigate to the **SOFTWARE\Microsoft\COM3\LocalComputerTable\ MyComputer** subkey. Click on the subkey to select it and display its values in the right-hand window.

4. Locate the **IMDBReservedBlobMemory** value entry. Use the DWORD Editor to change the IMDBReservedBlobMemory value to the desired size (in MB).

TIP: The default value of IMDBReservedBlobMemory is 512.

Setting the Default COM+ Authorization Level

You can set the default authorization level for use with DCOM by following these steps:

1. Launch Regedt32.

2. Select the Window menu option for **HKEY_LOCAL_MACHINE**.

3. Use the tree control in the left-hand window to navigate to the **SOFTWARE\Microsoft\COM3\LocalComputerTable\ MyComputer** subkey. Click on the subkey to select it and display its values in the right-hand window.

4. Locate the **DefaultAuthenticationLevel** value entry. Use the DWORD Editor to change the DefaultAuthenticationLevel value to the desired level of DCOM authentication unless overridden.

TIP: The default value of DefaultAuthenticationLevel is 4.

Setting the Default COM+ Impersonation Level

You can set the default impersonation level for use with DCOM by following these steps:

1. Launch Regedt32.

2. Select the Window menu option for **HKEY_LOCAL_MACHINE**.

3. Use the tree control in the left-hand window to navigate to the **SOFTWARE\Microsoft\COM3\LocalComputerTable\ MyComputer** subkey. Click on the subkey to select it and display its values in the right-hand window.

4. Locate the **DefaultImpersonationLevel** value entry. Use the DWORD Editor to change the DefaultImpersonationLevel value to the desired level of DCOM impersonation unless overridden.

TIP: The default value of DefaultImpersonationLevel is 2.

16. COM+

Determining the Location of COM+ Packages

The Registry details where COM+ keeps its package information for COM+ applications. If COM+ starts misbehaving, the Registry is one of the first places you should check:

1. Launch Regedt32.

2. Select the Window menu option for **HKEY_LOCAL_MACHINE**.

3. Use the tree control in the left-hand window to navigate to the **SOFTWARE\Microsoft\COM3\LocalComputerTable\ MyComputer** subkey. Click on the subkey to select it and display its values in the right-hand window.

4. Locate the **PackageInstallPath** value entry. If PackageInstallPath is empty, then COM+ is putting its information where it was installed. Otherwise, check to ensure that this value indeed points to a valid location and that COM+ data is actually there.

Determining the Location of Remote COM+ Packages

The Registry controls where COM+ keeps its package information for COM+ remote applications. If COM+ starts misbehaving, the Registry also is one of the first places administrators should check:

1. Launch Regedt32.

2. Select the Window menu option for **HKEY_LOCAL_MACHINE**.

3. Use the tree control in the left-hand window to navigate to the **SOFTWARE\Microsoft\COM3\LocalComputerTable\ MyComputer** subkey. Click on the subkey to select it and display its values in the right-hand window.

4. Locate the **RemotePackageInstallPath** value entry. If RemotePackageInstallPath is empty, then COM+ is putting its information where it was installed. Otherwise, check to ensure that this value indeed points to a valid location and that COM+ data is actually there.

16. COM+

Enabling a Computer to Function as a COM+ Router

The Registry controls whether a given COM+ installation can function as a COM+ router (performing load balancing between local and remote COM+ computers). Administrators can enable this capability as follows:

1. Launch Regedt32.

2. Select the Window menu option for **HKEY_LOCAL_MACHINE**.

3. Use the tree control in the left-hand window to navigate to the **SOFTWARE\Microsoft\COM3\LocalComputerTable\ MyComputer** subkey. Click on the subkey to select it and display its values in the right-hand window.

4. Locate the **IsRouter** value entry. Use the String Editor to change its value to Y to make the current COM+ installation into a COM+ router.

Enabling COM+ Resource Pooling

The Registry controls the ability to pool (keep in memory after use) high-cost items like ODBC connections. To turn on this feature manually, administrators can manipulate the Registry as follows:

1. Launch Regedt32.

2. Select the Window menu option for **HKEY_LOCAL_MACHINE**.

3. Use the tree control in the left-hand window to navigate to the **SOFTWARE\Microsoft\COM3\LocalComputerTable\ MyComputer** subkey. Click on the subkey to select it and display its values in the right-hand window.

4. Locate the **ResourcePoolingEnabled** value entry. Use the String Editor to change the ResourcePoolingEnabled value to Y to turn on COM+ Resource Pooling.

16. COM+

Preventing COM+ Transaction Timeout Failures

The Registry controls how long COM+ waits for a transaction to be signaled as complete or aborted. In cases in which network delays might be significant, you might need to increase this length of time. Administrators can access this vital setting via the Registry as follows:

1. Launch Regedt32.

2. Select the Window menu option for **HKEY_LOCAL_MACHINE**.

3. Use the tree control in the left-hand window to navigate to the **SOFTWARE\Microsoft\COM3\LocalComputerTable\ MyComputer** subkey. Click on the subkey to select it and display its values in the right-hand window.

4. Locate the **TransactionTimeout** value entry. Use the DWORD Editor to change the TransactionTimeout value to the time (in seconds) desired.

TIP: *The default value of TransactionTimeout is 60.*

---

16. COM+

Chapter 17

Internet Information Server

(continued)

In Brief

IIS stands for Internet Information Server, a combination of a number of Internet-oriented server applications bundled by Microsoft into one product. These applications include FTP, HTTP (WWW), and Index Server (which supports word-based online document searches). All the IIS server applications use the Registry for storing and manipulating vital aspects of their behavior, such as threading, caching, and security.

Protocol Support

With IIS, an administrator can provide a complete Internet solution to clients. IIS supports the following Internet RFC (Request for Comments):

- *File Transfer Protocol (FTP)*—Allows remote downloading and uploading of files (both text and binary).

- *Hypertext Transfer Protocol (HTTP)*—Provides complete Web site capabilities, including Active Server Pages (ASP).

- *Secure HTTP (HTTPS)*—Provides complete Web site capabilities, including Active Server Pages.

- *Secure Sockets Layer (SSL)*—Allows e-commerce without fear of stolen credit card numbers or other sensitive information.

Caching

Most of IIS's services take advantage of in-memory caching of their data to speed performance. The cache settings are controlled via the Registry. For example, suppose a Web page is often requested by clients, but lives across a rather slow network connection. By caching this Web page, IIS can satisfy user requests far more quickly because it is already in memory on the server computer. Similarly, an often-requested large FTP file can be cached (stored in memory) and provided to clients very fast, because it does not have to be located on disk each time it is requested.

17. Internet Information Server

Threading

IIS runs on Windows NT/2000 Server, so it can take advantage of symmetric multiprocessing (SMP) and maintain a per-processor thread pool to process incoming service requests. (All 32-bit Windows operating systems are multithreaded, of course, but only NT/2000 Server supports SMP.) A group of Registry entries controls this thread pool.

Security

IIS supports the Windows NT security model, which has carried over to Windows 2000 and has support for guest users in several flavors:

- *Anonymous FTP*—IIS FTP service has the option to permit users to log on with the generic name "anonymous". A set of Registry entries keeps track of the account used for such anonymous users when they log in and out, as well as the logins of non-anonymous users.

- *Anonymous HTTP*—Anonymous HTTP is the norm (that is, no login information is sent along with an HTTP request), but you can use the Registry to turn off anonymous HTTP for security reasons.

- *Guest logins*—Windows NT/2000's security model provides a guest account (its user name is "Guest" and it has an empty password) usable by anyone. This account is disabled by default; however, you must be careful when manipulating the Registry for some IIS services, because they can be set to allow IIS clients to obtain NT/2000 user privileges with the Guest account.

Immediate Solutions

Enabling IIS Server Memory Caching

IIS can cache varying amounts of its working data to speed responses to information requests. To enable this feature, tweak the Registry as follows:

1. Launch Regedt32.

2. Select the Window menu option for **HKEY_LOCAL_MACHINE**.

3. Use the tree control in the left-hand window to navigate to the **SYSTEM\CurrentControlSet\Services\InetInfo\Parameters** subkey. Click on the subkey to select it and display its values in the right-hand window.

4. Locate the **DisableMemoryCache** value entry. Use the DWORD Editor to change the DisableMemoryCache value to 0 to enable IIS memory caching.

TIP: If the DisableMemoryCache value does not exist, you can create it with the Edit\Add value.

WARNING! You can set the DisableMemoryCache value only via the Registry; Internet Service Manager cannot change it.

Increasing IIS Server Queued Connections Maximum

IIS allows queuing of connections to speed performance. In some situations in which the default setting does not take full advantage of the hardware and resources of a given server, you can use the following steps to tweak the Registry to increase the connection queue maximum value:

1. Launch Regedt32.

2. Select the Window menu option for **HKEY_LOCAL_MACHINE**.

3. Use the tree control in the left-hand window to navigate to the **SYSTEM\CurrentControlSet\Services\InetInfo\Parameters** subkey. Click on the subkey to select it and display its values in the right-hand window.

4. Locate the **ListenBackLog** value entry. Use the DWORD Editor to change the ListenBackLog value to a higher number (the default is 25) to increase the number of permissible queued IIS connections.

TIP: *If the ListenBackLog value does not exist, you can create it with the Edit/Add value.*

WARNING! The ListenBackLog value has no upper limit, so change it with caution.

Increasing IIS Log File Update Frequency

IIS keeps its log file in memory until a certain size is reached, after which IIS writes the log file to disk. You can decrease the amount of RAM the log file uses and also force more frequent log file updates by changing the Registry as follows:

1. Launch Regedt32.

2. Select the Window menu option for **HKEY_LOCAL_MACHINE**.

3. Use the tree control in the left-hand window to navigate to the **SYSTEM\CurrentControlSet\Services\InetInfo\Parameters** subkey. Click on the subkey to select it and display its values in the right-hand window.

4. Locate the **LogFileBatchSize** value entry. Use the DWORD Editor to change its value to a lower number of kilobytes to force more frequent batch file updates.

TIP: *The range for the LogFileBatchSize value is from 1 through **0xFFFFFFFF**, with a default of 64. If the LogFileBatchSize value does not exist, you can create it with the Edit/Add value.*

Adjusting Per-Processor IIS Maximum Thread Execution During I/O Blocks

IIS can permit multiple threads to run on all processors of the machine while waiting on I/O. This allows information processing to go on while lengthy network operations complete. Normally, IIS uses an internal algorithm to determine this value, but you might want to increase or decrease the thread concurrency under these circumstances. To do this, change the Registry as follows:

1. Launch Regedt32.

2. Select the Window menu option for **HKEY_LOCAL_MACHINE**.

3. Use the tree control in the left-hand window to navigate to the **SYSTEM\CurrentControlSet\Services\InetInfo\Parameters** subkey. Click on the subkey to select it and display its values in the right-hand window.

4. Locate the **MaxConcurrency** value entry. Use the DWORD Editor to change the MaxConcurrency value to a hard-coded number of allowable concurrent threads or set the MaxConcurrency to 0 to allow the system to determine the number of threads using an operating system internal algorithm.

*TIP: The range for the MaxConcurrency value is from 0 through **0xFFFFFFFF**, with a default of 0. If the MaxConcurrency value does not exist, you can create it with the Edit|Add value.*

Increasing the Per-Processor IIS Thread Pool

IIS maintains a pool of threads that watch for network requests. When a request arrives, IIS obtains the request from the network and begins processing. This is a per-processor pool and has a default setting created by IIS at installation. You might want to increase or decrease this value to optimize performance. To do this, change the Registry as follows:

1. Launch Regedt32.

2. Select the Window menu option for **HKEY_LOCAL_MACHINE**.

17. Internet Information Server

3. Use the tree control in the left-hand window to navigate to the **SYSTEM\CurrentControlSet\Services\InetInfo\Parameters** subkey. Click on the subkey to select it and display its values in the right-hand window.

4. Locate the **MaxPoolThreads** value entry. Use the DWORD Editor to change the MaxPoolThreads value to the number of threads you want to have in the pool watching for network requests.

TIP: *The range for the MaxPoolThreads value is from 0 through **0xFFFFFFFF**, with a default of 10. If the MaxPoolThreads value does not exist, you can create it with the Edit|Add value.*

WARNING! The MaxPoolThreads value should not exceed 20 due to context-switching overhead.

Increasing RAM Allocated for IIS Memory Caching

IIS can cache much of its working data sets in memory to speed performance. After you have turned on this feature, you must decide how much RAM to allocate to IIS memory caching and change the appropriate Registry setting as follows:

1. Launch Regedt32.

2. Select the Window menu option for **HKEY_LOCAL_MACHINE**.

3. Use the tree control in the left-hand window to navigate to the **SYSTEM\CurrentControlSet\Services\InetInfo\Parameters** subkey. Click on the subkey to select it and display its values in the right-hand window.

4. Locate the **MemoryCacheSize** value entry. Use the DWORD Editor to change the MemoryCacheSize value to the desired RAM usage level for optimal performance.

TIP: *The range for the MemoryCacheSize value is from 0 through **0xFFFFFFFF**, with a default of 3,072,000 (bytes or 3MB). If the MemoryCacheSize value does not exist, you can create it with the Edit|Add value.*

---

17. Internet Information Server

Preventing Slow Connection IIS File Transfer Timeouts

IIS has a set file transmission timeout value (in bytes sent during a fixed interval) to detect failed file transfers over a network. In some intranets, this setting can cause inappropriate file transfer timeout failures. You can reset the bytes-sent value to a lower value and fix the problem by tweaking the Registry as follows:

1. Launch Regedt32.

2. Select the Window menu option for **HKEY_LOCAL_MACHINE**.

3. Use the tree control in the left-hand window to navigate to the **SYSTEM\CurrentControlSet\Services\InetInfo\Parameters** subkey. Click on the subkey to select it and display its values in the right-hand window.

4. Locate the **MinFileKbSec** value entry. Use the DWORD Editor to change the MinFileKbSec value to a value appropriate for your network environment.

TIP: *The range for the MinFileKbSec value is from 1 through 8,192, with a default of 1,000. If the MinFileKbSec value does not exist, you can create it with the Edit|Add value.*

WARNING! *Despite the "Kb" in the value entry's name, the setting is in bytes, not kilobytes.*

Adjusting IIS TTL Cache Settings

The IIS cache manager includes code to "scavenge" the cache, removing objects that have not had their data referenced (read or changed) for a set number of seconds (this is called *time to live*, or *TTL*). If a system has a very volatile data set, this value might need to be adjusted to a lower value. Conversely, if the data in the cache needs to account for slow network traffic, the value might need to be adjusted upward. Either way, you can use the Registry to change TTL cache settings. To do this, follow these steps:

1. Launch Regedt32.

2. Select the Window menu option for **HKEY_LOCAL_MACHINE**.

3. Use the tree control in the left-hand window to navigate to the **SYSTEM\CurrentControlSet\Services\InetInfo\Parameters** subkey. Click on the subkey to select it and display its values in the right-hand window.

4. Locate the **ObjectCacheTTL** value entry. Use the DWORD Editor to change the ObjectCacheTTL value to set it to a lower value in seconds to promote faster cache reuse or a higher value in seconds to allow longer cache lifetimes.

TIP: The ObjectCacheTTL value has a default of 30. If the ObjectCacheTTL value does not exist, you can create it with the Edit/Add value.

WARNING! Setting the ObjectCacheTTL value to 0xFFFFFFFF disables the cache scavenger.

Adjusting IIS Thread TTL Settings

IIS normally maintains the threads in its pool even when there is no I/O activity. After a set period, however, it releases threads on the assumption that they are not needed. You can adjust the time threads live while idle by tweaking the Registry as follows:

1. Launch Regedt32.

2. Select the Window menu option for **HKEY_LOCAL_MACHINE**.

3. Use the tree control in the left-hand window to navigate to the **SYSTEM\CurrentControlSet\Services\InetInfo\Parameters** subkey. Click on the subkey to select it and display its values in the right-hand window.

4. Locate the **ThreadTimeout** value entry. Use the DWORD Editor to change the ThreadTimeout value to a number in seconds that an IIS thread is allowed to live while idle.

TIP: The ThreadTimeout value has a default of 86,400 (seconds or 24 hours). If the ThreadTimeout value does not exist, you can create it with the Edit/Add value.

Adjusting IIS Anonymous User Security Token TTL Settings

IIS can permit anonymous users for its various services. To allow this, IIS creates a user security token for the anonymous user and caches it so that the internal Windows 2000 logon process does not have to be repeated for every anonymous login. This token has a preset TTL in the cache before it is scavenged due to its not being used. You can change this value to optimize system performance or prevent security breaches (which can happen when a rogue process obtains this token illegally). To do this, follow these steps:

1. Launch Regedt32.

2. Select the Window menu option for **HKEY_LOCAL_MACHINE**.

3. Use the tree control in the left-hand window to navigate to the **SYSTEM\CurrentControlSet\Services\InetInfo\Parameters** subkey. Click on the subkey to select it and display its values in the right-hand window.

4. Locate the **UserTokenTTL** value entry. Use the DWORD Editor to change the UserTokenTTL value to the desired time in seconds for the anonymous user security ID token to remain in the cache.

TIP: The UserTokenTTL value has a default of 900 seconds (15 minutes). If the UserTokenTTL value does not exist, you can create it with the Edit\Add value.

Permitting Anonymous Users for Specific IIS Services

You can configure acceptance or rejection of anonymous logins for any of the IIS services by tweaking the Registry as follows:

1. Launch Regedt32.

2. Select the Window menu option for **HKEY_LOCAL_MACHINE**.

3. Use the tree control in the left-hand window to navigate to the **SYSTEM\CurrentControlSet\Services***[Service Name]***\Parameters** subkey, where *[Service Name]* is

MSFTPSVC for FTP, or W3SVC for WWW/HTTP service. Click on the subkey to select it and display its values in the right-hand window.

4. Locate the **AllowAnonymous** value entry. Use the DWORD Editor to change the AllowAnonymous value to 1 to permit anonymous logins and 0 to disable them.

TIP: *If the AllowAnonymous value does not exist, you can create it with the Edit/Add value.*

Checking the Actual Username for Anonymous IIS Logins

If you are going to permit anonymous logins for a given IIS service, you must also configure (using IIS) the username to be used. If problems occur with anonymous logins, you can check the Registry to make sure the appropriate entries are not corrupted. To check this value, follow these steps:

1. Launch Regedt32.

2. Select the Window menu option for **HKEY_LOCAL_MACHINE**.

3. Use the tree control in the left-hand window to navigate to the **SYSTEM\CurrentControlSet\Services\\[*Service Name]*\Parameters** subkey, where *[Service Name]* is MSFTPSVC for FTP, or W3SVC for WWW/HTTP service. Click on the subkey to select it and display its values in the right-hand window.

4. Locate the **AnonymousUserName** value entry. Use the String Editor to make sure AnonymousUserName is set to the desired value.

WARNING! *You must set the AnonymousUserName value via Internet Service Manager so you can also set its password.*

Preventing Slow IIS Connection Timeouts

IIS has a preset timeout value for connections with no activity. In rare cases, you might need to increase this value to prevent invalid timeouts on a slow intranet. You can fix such a problem by adjusting the Registry as follows:

1. Launch Regedt32.

2. Select the Window menu option for **HKEY_LOCAL_MACHINE**.

3. Use the tree control in the left-hand window to navigate to the **SYSTEM\CurrentControlSet\Services\[*Service Name]*\Parameters** subkey, where *[Service Name]* is MSFTPSVC for FTP, or W3SVC for WWW/HTTP service. Click on the subkey to select it and display its values in the right-hand window.

4. Locate the **ConnectionTimeOut** value entry. Use the DWORD Editor to change the ConnectionTimeOut value to a setting more suited for the current network conditions.

TIP: *The default for the ConnectionTimeOut value is 600 seconds (10 minutes).*

Setting the Log File Path for IIS Logs

You can control a number of log file behaviors via Registry keys. For example, you can specify the directory in which log files are written by following these steps:

1. Launch Regedt32.

2. Select the Window menu option for **HKEY_LOCAL_MACHINE**.

3. Use the tree control in the left-hand window to navigate to the **SYSTEM\CurrentControlSet\Services\[*Service Name]*\Parameters** subkey, where *[Service Name]* is MSFTPSVC for FTP, or W3SVC for WWW/HTTP service. Click on the subkey to select it and display its values in the right-hand window.

4. Locate the **LogFileDirectory** value entry. Use the String Editor to change the LogFileDirectory value to specify the desired directory for log file creation.

17. Internet Information Server

Setting the Log File Format for IIS Logs

IIS services write log files of various events in their operation. Administrators can control a number of log file behaviors via Registry keys, including whether the log uses NCSA (National Center For Supercomputing Applications) format, in which many Internet implementations were created (including the first Web browser). To specify whether a log uses NCSA format, follow these steps:

1. Launch Regedt32.

2. Select the Window menu option for **HKEY_LOCAL_MACHINE**.

3. Use the tree control in the left-hand window to navigate to the **SYSTEM\CurrentControlSet\Services***[Service Name]***\Parameters** subkey, where *[Service Name]* is MSFTPSVC for FTP, or W3SVC for WWW/HTTP service. Click on the subkey to select it and display its values in the right-hand window.

4. Locate the **LogFileFormat** value entry. Use the DWORD Editor to change the LogFileFormat value to 3 to use NCSA format or 0 for standard format.

TIP: *The default for the LogFileFormat value is 0. If the LogFileFormat value does not exist, you can create it with the Edit/Add value.*

Setting the New Log File Creation Interval for IIS Logs

You can control a number of log file behaviors via Registry keys, including whether log files are created daily, weekly, monthly, or only when the size limit is reached. To specify how frequently log files are created, follow these steps:

1. Launch Regedt32.

2. Select the Window menu option for **HKEY_LOCAL_MACHINE**.

3. Use the tree control in the left-hand window to navigate to the **SYSTEM\CurrentControlSet\Services***[Service Name]***\Parameters** subkey, where *[Service Name]* is

MSFTPSVC for FTP, or W3SVC for WWW/HTTP service. Click on the subkey to select it and display its values in the right-hand window.

4. Locate the **LogFilePeriod** value entry. Use the DWORD Editor to change the LogFilePeriod value to one of the settings shown in Table 17.1.

TIP: If the LogFilePeriod value does not exist, you can create it with the Edit/Add value.

Table 17.1   IIS service log file creation interval settings.

Value	Creation Interval
0	No interval; new log created when old log exceeds size limit
1	New log created daily
2	New log created weekly
3	New log created monthly

Preventing New IIS Log File Creation

You can control a number of log file behaviors via Registry keys, including the size limit. When you specify a size limit and the size limit is met, IIS is forced to create a new log file. To specify the log file's size limit, follow these steps:

1. Launch Regedt32.

2. Select the Window menu option for **HKEY_LOCAL_MACHINE**.

3. Use the tree control in the left-hand window to navigate to the **SYSTEM\CurrentControlSet\Services\[*Service Name*]\Parameters** subkey, where *[Service Name]* is MSFTPSVC for FTP, or W3SVC for WWW/HTTP service. Click on the subkey to select it and display its values in the right-hand window.

4. Locate the **LogFileTruncateSize** value entry. Use the DWORD Editor to change the LogFileTruncateSize value (to the size in bytes) that specifies at which point IIS should start a new log file.

17. Internet Information Server

TIP: *The default for the LogFileTruncateSize value is **0xFFFFFFFF**. If the LogFileTruncateSize value does not exist, you can create it with the Edit/Add value.*

WARNING! Setting the LogFileTruncateSize value to 0 means the file can grow to an unlimited size.

Forcing IIS Logs to be Written to an ODBC Database

You can also control whether IIS writes the log file information to an ODBC (Open Database Connectivity) database rather than a file. To do this, follow these steps:

1. Launch Regedt32.

2. Select the Window menu option for **HKEY_LOCAL_MACHINE**.

3. Use the tree control in the left-hand window to navigate to the **SYSTEM\CurrentControlSet\Services\\[Service Name]\Parameters** subkey, where **[Service Name]** is MSFTPSVC for FTP, or W3SVC for WWW/HTTP service. Click on the subkey to select it and display its values in the right-hand window.

4. Locate the **LogType** value entry. Use the DWORD Editor to change the LogType value to one of the values shown in Table 17.2.

TIP: *If the LogType value does not exist, you can create it with the Edit/Add value.*

Table 17.2   IIS service log file creation type settings.

Value	Creation Type
0	No logging
1	Log to file
2	Log to ODBC database

Adjusting the Maximum Allowable Simultaneous IIS Connections

High-traffic networks can sometimes run into an unexpected problem when IIS starts refusing connections for no apparent reason. In this case, what is actually happening is that an internal limit for maximum simultaneous connections has been reached. The solution is to adjust the internal IIS maximum connections value in the Registry as follows:

1. Launch Regedt32.

2. Select the Window menu option for **HKEY_LOCAL_MACHINE**.

3. Use the tree control in the left-hand window to navigate to the **SYSTEM\CurrentControlSet\Services\[*Service Name]*\Parameters** subkey, where *[Service Name]* is MSFTPSVC for FTP, or W3SVC for WWW/HTTP service. Click on the subkey to select it and display its values in the right-hand window.

4. Locate the **MaxConnections** value entry. Use the DWORD Editor to change the MaxConnections value to a setting consistent with traffic needs of the network.

TIP: *The default for the MaxConnections value is 100,000. If the MaxConnections value does not exist, you can create it with the Edit/Add value.*

Changing an IIS Service Logon Comment Message

Some IIS services provide a connection message that is sent when a successful connection to them is made. You can customize the generic message by modifying the Registry as follows:

1. Launch Regedt32.

2. Select the Window menu option for **HKEY_LOCAL_MACHINE**.

3. Use the tree control in the left-hand window to navigate to the **SYSTEM\CurrentControlSet\Services\[*Service Name]*\Parameters** subkey, where *[Service Name]* is

MSFTPSVC for FTP, or W3SVC for WWW/HTTP service. Click on the subkey to select it and display its values in the right-hand window.

4. Locate the **ServerComment** value entry. Use the String Editor to change the ServerComment value to a custom welcome message.

TIP: *If the ServerComment value does not exist, you can create it with the Edit/Add value.*

Changing the WWW IIS Service "Access Denied" Message

Another Registry setting you can configure is the message that is sent when HTTP access is denied to a user. You can change the **AccessDeniedMessage** value to a simple HTML string that contains instructions for reporting the problem. To modify the AccessDeniedMessage value, follow these steps:

1. Launch Regedt32.

2. Select the Window menu option for **HKEY_LOCAL_MACHINE**.

3. Use the tree control in the left-hand window to navigate to the **SYSTEM\CurrentControlSet\Services\W3SVC\Parameters** subkey. Click on the subkey to select it and display its values in the right-hand window.

4. Locate the **AccessDeniedMessage** value entry. Use the String Editor to change the AccessDeniedMessage value to a custom message, such as HTML text that explains how to report the error.

TIP: *HTML text does not need line breaks or carriage returns.*

Enabling Guest Logons for WWW IIS Service

Another Registry setting you can configure pertains to whether WWW logons are allowed using the guest account. To access this Registry setting, follow these steps:

1. Launch Regedt32.

2. Select the Window menu option for **HKEY_LOCAL_MACHINE**.

3. Use the tree control in the left-hand window to navigate to the **SYSTEM\CurrentControlSet\Services\W3SVC\Parameters** subkey. Click on the subkey to select it and display its values in the right-hand window.

4. Locate the **AllowGuestAccess** value entry. Use the DWORD Editor to change the AllowGuestAccess value to 1 to enable guest logons and 0 to disable them.

TIP: The default for the AllowGuestAccess value is 1. If the AllowGuestAccess value does not exist, you can create it with the Edit/Add value.

WARNING! Guest logons normally have high-level access. You should enable them only when all possible users are considered appropriate for this type of access.

Enabling IIS ISAPI Extension Memory Caching

Another Registry setting you can configure controls whether ISAPI (Internet Service Application Programming Interface) extensions are cached rather than reloaded each time they are invoked. To access this Registry setting, follow these steps:

1. Launch Regedt32.

2. Select the Window menu option for **HKEY_LOCAL_MACHINE**.

3. Use the tree control in the left-hand window to navigate to the **SYSTEM\CurrentControlSet\Services\W3SVC\Parameters** subkey. Click on the subkey to select it and display its values in the right-hand window.

4. Locate the **CacheExtensions** value entry. Use the DWORD Editor to change the CacheExtensions value to 0 to disable caching and 1 to enable it.

TIP: The default for the CacheExtensions value is 1. If CacheExtensions does not exist, you can create it with the Edit/Add value.

WARNING! Turning off ISAPI extension caching can severely degrade system performance. It should be disabled only to locate a problem and then reactivated.

Preventing Proxy File Caching for IIS WWW Service Files

Another Registry setting you can configure controls the ability to disable file caching by proxy servers (preventing often-changed files from being inappropriately cached). To access this Registry setting, follow these steps:

1. Launch Regedt32.

2. Select the Window menu option for **HKEY_LOCAL_MACHINE**.

3. Use the tree control in the left-hand window to navigate to the **SYSTEM\CurrentControlSet\Services\W3SVC\Parameters** subkey. Click on the subkey to select it and display its values in the right-hand window.

4. Locate the **GlobalExpire** value entry. Use the DWORD Editor to change the GlobalExpire value to 0 to prevent proxy file caching.

TIP: If the GlobalExpire value does not exist, you can create it with the Edit/Add value.

Enabling Logging of Successful WWW IIS Service Requests

Another Registry key enables you to control whether successful HTTP requests are written to the log file. Enabling this setting allows tracking HTTP server traffic, but it can also generate much larger log files. To access this Registry setting, follow these steps:

1. Launch Regedt32.

2. Select the Window menu option for **HKEY_LOCAL_MACHINE**.

3. Use the tree control in the left-hand window to navigate to the **SYSTEM\CurrentControlSet\Services\W3SVC\Parameters** subkey. Click on the subkey to select it and display its values in the right-hand window.

4. Locate the **LogSuccessfulRequests** value entry. Use the DWORD Editor to change the LogSuccessfulRequests value to 1 to enable logging and 0 to disable it.

TIP: If the LogSuccessfulRequests value does not exist, you can create it with the Edit/Add value.

Enabling Logging of Unsuccessful WWW IIS Service Requests

Another Registry setting—LogErrorRequests—controls whether error-generating HTTP requests are written to the log file. Enabling LogErrorRequests allows you to track HTTP server traffic, but it can generate much larger log files. To access this Registry setting, follow these steps:

1. Launch Regedt32.

2. Select the Window menu option for **HKEY_LOCAL_MACHINE**.

3. Use the tree control in the left-hand window to navigate to the **SYSTEM\CurrentControlSet\Services\W3SVC\Parameters** subkey. Click on the subkey to select it and display its values in the right-hand window.

4. Locate the **LogErrorRequests** value entry. Use the DWORD Editor to change the LogErrorRequests value to 1 to enable logging and 0 to disable it.

TIP: *If the LogErrorRequests value does not exist, you can create it with the Edit/Add value.*

Enabling ODBC Database Connection Pooling for WWW IIS Service

Another Registry setting enables *IDC (Internet Database Connector)* and ODBC connection pooling and reuse to avoid the overhead of recreating the database connections. To access this Registry setting, follow these steps:

1. Launch Regedt32.

2. Select the Window menu option for **HKEY_LOCAL_MACHINE**.

3. Use the tree control in the left-hand window to navigate to the **SYSTEM\CurrentControlSet\Services\W3SVC\Parameters** subkey. Click on the subkey to select it and display its values in the right-hand window.

4. Locate the **PoolIDCConnections** value entry. Use the DWORD Editor to change the PoolIDCConnections value to 1 to enable connection pooling and 0 to disable it.

17. Internet Information Server

TIP: *If the PoolIDCConnections value does not exist, you can create it with the Edit/Add value.*

Preventing ODBC Connection Pooling Timeouts with WWW IIS Service

As described in the preceding Immediate Solution, you can configure the Registry to enable IDC and ODBC connection pooling and reuse to avoid the overhead of re-creating the database connections. The Registry also contains a setting that controls how long the connections are pooled before being closed. You might need to increase this setting for slow networks. To configure how long connections are pooled before being closed, follow these steps:

1. Launch Regedt32.

2. Select the Window menu option for **HKEY_LOCAL_MACHINE**.

3. Use the tree control in the left-hand window to navigate to the **SYSTEM\CurrentControlSet\Services\W3SVC\Parameters** subkey. Click on the subkey to select it and display its values in the right-hand window.

4. Locate the **PoolIDCConnectionsTimeout** value entry. Use the DWORD Editor to change the PoolIDCConnectionsTimeout value to a higher value in seconds to prevent pooling timeouts over slow networks.

TIP: *The default for the PoolIDCConnectionsTimeout value is 30 seconds. You must enable ODBC connection pooling before you can use this feature. If the PoolIDCConnectionsTimeout value does not exist, you can create it with the Edit/Add value.*

Preventing WWW IIS Service CGI Script Timeouts on Slow Networks

Another Registry setting determines how long IIS waits for a response from a CGI (Common Gateway Interface) script. You might need to adjust the wait time for slow networks. To do this, follow these steps:

1. Launch Regedt32.

2. Select the Window menu option for **HKEY_LOCAL_MACHINE**.

3. Use the tree control in the left-hand window to navigate to the **SYSTEM\CurrentControlSet\Services\W3SVC\Parameters** subkey. Click on the subkey to select it and display its values in the right-hand window.

4. Locate the **ScriptTimeout** value entry. Use the DWORD Editor to change the ScriptTimeout value to a higher value in milliseconds for slow networks.

TIP: The default for the ScriptTimeout value is 384; the range is from 1 through 80,000,000. If the ScriptTimeout value does not exist, you can create it with the Edit|Add value.

Setting the SSL Port for WWW IIS Service

Another Registry setting specifies the port used for SSL connections for e-commerce. To access this Registry setting, follow these steps:

1. Launch Regedt32.

2. Select the Window menu option for **HKEY_LOCAL_MACHINE**.

3. Use the tree control in the left-hand window to navigate to the **SYSTEM\CurrentControlSet\Services\W3SVC\Parameters** subkey. Click on the subkey to select it and display its values in the right-hand window.

4. Locate the **SecurePort** value entry. Use the DWORD Editor to change the SecurePort value to that of the desired SSL port.

---

*TIP: The default for the SecurePort value is **0x1BB**. If the SecurePort value does not exist, you can create it with the Edit|Add value.*

Disabling Guest Access to IIS FTP Service

Another Registry setting specifies whether FTP logons are allowed using the guest account. To access this Registry setting, follow these steps:

1. Launch Regedt32.

2. Select the Window menu option for **HKEY_LOCAL_MACHINE**.

17. Internet Information Server

3. Use the tree control in the left-hand window to navigate to the
   SYSTEM\CurrentControlSet\Services\FTPSVC\Parameters
   subkey. Click on the subkey to select it and display its values in
   the right-hand window.

4. Locate the **AllowGuestAccess** value entry. Use the DWORD
   Editor to change the AllowGuestAccess value to 1 to enable
   guest logons and 0 to disable them.

TIP: The default for the AllowGuestAccess value is 1. If AllowGuestAccess does not exist, you can create it with the Edit|Add value.

WARNING! Guest logons normally have high-level access. You should enable them only when all possible users are considered appropriate for this type of access.

Disabling IIS FTP Keep-Alive Negotiations

Another Registry setting controls whether the FTP service uses the
keep-alive negotiation system. To access this Registry setting, follow
these steps:

1. Launch Regedt32.

2. Select the Window menu option for **HKEY_LOCAL_MACHINE**.

3. Use the tree control in the left-hand window to navigate to the
   SYSTEM\CurrentControlSet\Services\FTPSVC\Parameters
   subkey. Click on the subkey to select it and display its values in
   the right-hand window.

4. Locate the **AllowKeepAlives** value entry. Use the DWORD
   Editor to change the AllowKeepAlives value to 0 to disable FTP
   keep-alive negotiations and 1 to enable them.

TIP: The default for the AllowKeepAlives value is 1.

WARNING! Turning off keep-alive negotiation could significantly impact FTP server performance negatively.

Changing the IIS FTP Service Greeting Message

Another Registry setting specifies the greeting sent when users successfully log on to the FTP service. To specify a greeting for FTP users, follow these steps:

1. Launch Regedt32.

2. Select the Window menu option for **HKEY_LOCAL_MACHINE**.

3. Use the tree control in the left-hand window to navigate to the **SYSTEM\CurrentControlSet\Services\FTPSVC\Parameters** subkey. Click on the subkey to select it and display its values in the right-hand window.

4. Locate the **GreetingMessage** value entry. Use the String Editor to change the GreetingMessage value to a custom message appropriate for your organization.

Changing the IIS FTP Service Exit Message

Another Registry setting specifies the message sent when users successfully log off of the FTP service. To configure a logoff message for FTP users, follow these steps:

1. Launch Regedt32.

2. Select the Window menu option for **HKEY_LOCAL_MACHINE**.

3. Use the tree control in the left-hand window to navigate to the **SYSTEM\CurrentControlSet\Services\FTPSVC\Parameters** subkey. Click on the subkey to select it and display its values in the right-hand window.

4. Locate the **ExitMessage** value entry. Use the String Editor to change the ExitMessage value to a custom message appropriate for your organization.

17. Internet Information Server

Enabling Logging of Nonanonymous IIS FTP Logons

Another Registry setting controls whether to log nonanonymous logons to the event log for the FTP services. To access this Registry setting, follow these steps:

1. Launch Regedt32.

2. Select the Window menu option for **HKEY_LOCAL_MACHINE**.

3. Use the tree control in the left-hand window to navigate to the **SYSTEM\CurrentControlSet\Services\FTPSVC\Parameters** subkey. Click on the subkey to select it and display its values in the right-hand window.

4. Locate the **LogNonAnonymous** value entry. Use the DWORD Editor to change the LogNonAnonymous value to 1 to enable nonanonymous logons logging and 0 to disable it.

TIP: The default for the LogNonAnonymous value is 1.

Enabling Lowercase File Comparisons in IIS FTP Operations

Another Registry setting controls whether you are able to use lower-case as well as uppercase characters for FTP file comparisons. To access this Registry setting, follow these steps:

1. Launch Regedt32.

2. Select the Window menu option for **HKEY_LOCAL_MACHINE**.

3. Use the tree control in the left-hand window to navigate to the **SYSTEM\CurrentControlSet\Services\FTPSVC\Parameters** subkey. Click on the subkey to select it and display its values in the right-hand window.

4. Locate the **LowercaseFiles** value entry. Use the DWORD Editor to change the LowercaseFiles value to 1 to enable lowercase file name comparisons and 0 to disable it.

TIP: The default for the LowercaseFiles value is 0.

17. Internet Information Server

Changing the IIS FTP Service "Too Many Connections" Message

Another Registry setting specifies the message sent when the FTP service refuses a connection due to having reached its current connections maximum. To change this message, follow these steps:

1. Launch Regedt32.

2. Select the Window menu option for **HKEY_LOCAL_MACHINE**.

3. Use the tree control in the left-hand window to navigate to the **SYSTEM\CurrentControlSet\Services\FTPSVC\Parameters** subkey. Click on the subkey to select it and display its values in the right-hand window.

4. Locate the **MaxClientsMessage** value entry. Use the String Editor to change the MaxClientsMessage value to a message that is appropriate for your organization.

Enabling MS-DOS Style IIS FTP Service Directory Output

Another Registry setting controls whether to use an MS-DOS or Unix format for output of directories to FTP clients. To access this Registry setting, follow these steps:

1. Launch Regedt32.

2. Select the Window menu option for **HKEY_LOCAL_MACHINE**.

3. Use the tree control in the left-hand window to navigate to the **SYSTEM\CurrentControlSet\Services\FTPSVC\Parameters** subkey. Click on the subkey to select it and display its values in the right-hand window.

4. Locate the **MsdosDirOutput** value entry. Use the DWORD Editor to change the MsdosDirOutput value to 1 to enable MS-DOS–style directory listings and 0 to enable Unix-style directory listings.

TIP: The default for the MsdosDirOutput value is 1.

Enabling IIS FTP Service Read Aheads

Another Registry setting holds the number of bytes read into buffers by the FTP service prior to turning control of the connection over to the FTP client application. To access this Registry value, follow these steps:

1. Launch Regedt32.

2. Select the Window menu option for **HKEY_LOCAL_MACHINE**.

3. Use the tree control in the left-hand window to navigate to the **SYSTEM\CurrentControlSet\Services\FTPSVC\Parameters** subkey. Click on the subkey to select it and display its values in the right-hand window.

4. Locate the **UploadReadAhead** value entry. Use the DWORD Editor to change the UploadReadAhead value to the number of bytes read before control is sent to the application.

TIP: The default for the UploadReadAhead value is 48,000; the range is from 0 through 80,000,000.

Chapter 18

Internet Explorer 4+

(continued)

In Brief

Windows 2000 is fully configured to use Internet Explorer (IE) version 5 and the Active Desktop. IE5 supports all of IE4's functionality, and adds some new features of its own. There are a number of very useful Registry settings connected with IE4+ and Active Desktop, including personal directories and Start menu settings, IE security, IE user interface settings, and Active Desktop settings.

Personal Directories

Each Internet Explorer user has a set of personal directories that the Registry tracks. These directories allow each active user to have his or her own set of folders for storing documents, favorites, cookies (small text files used by IE). Using these folders avoids the complications that arise when users create their own folders with names that standardized applications do not recognize. Internet Explorer's personal directories include the following:

- My Documents folder
- History folder
- Favorites folder
- Startup folder (all programs executed when the shell starts)
- Recent Files folder
- IE Download Cache folder
- IE Cookies folder

IE Security

Although you can manage many security features using IE's user interface, IE stores some security settings per user in the Registry. Security features stored in the Registry include general security warning levels (such as when active content is about to be downloaded and run) and certificate verification using Authenticode. The Registry also controls the ability of IE to display images and videos, and to use the integrated script debugger.

IE User Interface

By tweaking the Registry, a Windows 2000 administrator can control many features of the IE4+ user interface, including toolbar and status bar display, URL (Universal Resource Locator) display (with or without full URL information in the status bar), and color schemes. In addition, administrators can configure the default start page and search engine URL via the Registry. Remember that while many of these settings can be set via the IE4+ graphical user interface (GUI), setting them via the Registry can save an administrator a lot of time and effort.

Active Desktop

The IE4+ Active Desktop combines the basic Windows 2000 desktop display with the power of HTML (hypertext markup language). This results in a new level of functionality, but also introduces some complexity for per-user administration. New elements that administrators must consider include the following:

- *Active Desktop wallpaper*—The Active Desktop supports graphic formats other than BMP (bitmap) for its background wallpaper, including HTML pages. The location of the wallpaper file is stored in the Registry.

- *URL taskbar*—An available Active Desktop toolbar contains the Address bar from IE or Explorer. It can be made unavailable to users via a Registry setting.

- *Channel support*—Channels are an attempt to bring standardized content to Web sites and have specialized support in Active Desktop. The Registry permits setting the default Channel brought up by the taskbar or IE/Explorer to one different from Microsoft's default setting.

Immediate Solutions

Forcing Automatic Reload of Remote URLs

One of the more powerful features of IE4+ is caching—maintaining a store of previously downloaded Web page components, so that they do not have to be downloaded again for each display. In cases where page content changes constantly, having a cache can require a manual refresh of each viewing. To force IE to automatically reload all remote URLs (which can greatly slow performance in some cases), you need to modify each user's Registry as follows:

1. Launch Regedt32.

2. Select the Window menu option for **HKEY_CURRENT_USER**.

3. Use the tree control in the left-hand window to navigate to the **SOFTWARE\Microsoft\Internet Explorer\Main** subkey. Click on the subkey to select it and display its values in the right-hand window.

4. Locate the **Cache_Update_Frequency** value entry. Use the String Editor to change the Cache_Update_Frequency value to **Once_Per_Page** to force automatic remote URL reloading. The default is once per session.

Setting the Channels Home Page

IE4+ has the option to support *channels*, specialized Web pages containing large amounts of focused content that can be automatically updated according to a predefined schedule. IE comes with a default selection of channels and a home page set to Microsoft's Channels site. You can disable the default channels home page for a user and set the Active Desktop default channels to another page. To do so, change the Registry as follows:

1. Launch Regedt32.

2. Select the Window menu option for **HKEY_CURRENT_USER**.

3. Use the tree control in the left-hand window to navigate to the **SOFTWARE\Microsoft\Internet Explorer\Main** subkey. Click on the subkey to select it and display its values in the right-hand window.

18. Internet Explorer 4+

4. Locate the **ChannelsURL** value entry. Use the String Editor to change the ChannelsURL value to specify the URL for the desired channels home page for the user.

5. Locate the **ChannelsFirstURL** value entry. Use the String Editor to change the ChannelsFirstURL value to specify the desired first channel page displayed for the user (from those available in the previous step's home page).

Disabling the Scripting Debugger

IE4+ supports a scripting debugger for finding problems in ActiveX scripts in Web pages. In some situations this could pose a security hazard, such as allowing an unauthorized user to see scripting code that checks passwords. To disable this feature on a per-user basis, you can modify the Registry as follows:

1. Launch Regedt32.

2. Select the Window menu option for **HKEY_CURRENT_USER**.

3. Use the tree control in the left-hand window to navigate to the **SOFTWARE\Microsoft\Internet Explorer\Main** subkey. Click on the subkey to select it and display its values in the right-hand window.

4. Locate the **Disable Script Debugger** value entry. Use the String Editor to change the Disable Script Debugger value to Yes to prevent an unauthorized user from accessing HTML scripts via the debugger when a page crashes.

TIP: *You should disable access to the View|Source menu option in IE4+ if examining HTML scripts poses a security hazard.*

Disabling Image Display in HTML Pages

Many users viewing graphics-intensive Web pages can significantly impact some networks. You can tweak the Registry to turn off image display for a user who does not need to be able to view images. To do so, follow these steps:

1. Launch Regedt32.

2. Select the Window menu option for **HKEY_CURRENT_USER**.

3. Use the tree control in the left-hand window to navigate to the **SOFTWARE\Microsoft\Internet Explorer\Main** subkey. Click on the subkey to select it and display its values in the right-hand window.

4. Locate the **Display Inline Images** value entry. Use the String Editor to change the Display Inline Images value to No to prevent images from displaying on HTML pages viewed by the user.

WARNING! Setting the Display Inline Images value to No disables all images on HTML pages.

Disabling Video Display in HTML Pages

Many users viewing video-intensive Web pages can significantly impact some networks. You can tweak the Registry to turn off video display for a user who does not need video capabilities. To do so, follow these steps:

1. Launch Regedt32.

2. Select the Window menu option for **HKEY_CURRENT_USER**.

3. Use the tree control in the left-hand window to navigate to the **SOFTWARE\Microsoft\Internet Explorer\Main** subkey. Click on the subkey to select it and display its values in the right-hand window.

4. Locate the **Display Inline Videos** value entry. Use the String Editor to change the Display Inline Videos value to No to disable videos in HTML pages viewed by the user.

WARNING! Setting the Display Inline Videos value to No disables all videos in HTML pages.

Setting the Default HTML Page

Whenever an attempt to load a URL fails, IE4+ displays a default HTML page that is set to one of Microsoft's devising during installation. You can reassign the default page to display another, more appropriate

18. Internet Explorer 4+

page (such as a page containing a URL to send mail to the network administrator) by editing the Registry for each user who needs the change. To implement this change, follow these steps:

1. Launch Regedt32.

2. Select the Window menu option for **HKEY_CURRENT_USER**.

3. Use the tree control in the left-hand window to navigate to the **SOFTWARE\Microsoft\Internet Explorer\Main** subkey. Click on the subkey to select it and display its values in the right-hand window.

4. Locate the **Local Page** value entry. Use the String Editor to change the Local Page value to the full path of the desired failover HTML page.

Disabling Background Sound Playback in HTML Pages

Many users viewing sound-intensive Web pages (such as pages with complex background sounds) can significantly impact some networks. If a user does not need to use background sound for their authorized work, you can turn off the background sound playback by tweaking the Registry as follows:

1. Launch Regedt32.

2. Select the Window menu option for **HKEY_CURRENT_USER**.

3. Use the tree control in the left-hand window to navigate to the **SOFTWARE\Microsoft\Internet Explorer\Main** subkey. Click on the subkey to select it and display its values in the right-hand window.

4. Locate the **Play_Background_Sounds** value entry. Use the String Editor to change the Play_Background_Sounds value to No to prevent any background sounds in HTML pages being played for the user.

WARNING! Turning off the Play_Background_Sounds value disables all background sounds in HTML pages.

Enabling History Persistence on Exit

When a user exits the Web browser, IE4+ has the option to save the user's history (list of visited URLs) to disk. Normally, this option is turned off in Windows 2000 Server to save disk space, but if you need to enable this feature for a user, you can adjust the Registry as follows:

1. Launch Regedt32.

2. Select the Window menu option for **HKEY_CURRENT_USER**.

3. Use the tree control in the left-hand window to navigate to the **SOFTWARE\Microsoft\Internet Explorer\Main** subkey. Click on the subkey to select it and display its values in the right-hand window.

4. Locate the **Save_Session_History_On_Exit** value entry. Use the String Editor to change the Save_Session_History_On_Exit value to Yes to enable persisting History folder entries.

Setting the Default Search Engine

When a user clicks on the IE4+ Search button, the Search window appears within IE. The contents of the Search window (that is, the search engine interface that displays) is controlled on a per-user basis by the Registry. To access the Registry key that specifies the contents of the Search window, follow these steps:

1. Launch Regedt32.

2. Select the Window menu option for **HKEY_CURRENT_USER**.

3. Use the tree control in the left-hand window to navigate to the **SOFTWARE\Microsoft\Internet Explorer\Main** subkey. Click on the subkey to select it and display its values in the right-hand window.

4. Locate the **Search Page** value entry. Use the String Editor to change the Search Page value to the URL of the desired local or remote search engine.

Disabling Full URL Display

IE4+ can display the full URL of each new page in its window caption. In some cases, displaying a complete URL could be a security breach (such as displaying a hidden directory name). To disable this feature, you can change the Registry on a per-user basis by following these steps:

1. Launch Regedt32.

2. Select the Window menu option for **HKEY_CURRENT_USER**.

3. Use the tree control in the left-hand window to navigate to the **SOFTWARE\Microsoft\Internet Explorer\Main** subkey. Click on the subkey to select it and display its values in the right-hand window.

4. Locate the **Show_FullURL** value entry. Use the String Editor to change the Show_FullURL value to No to disable full URL display.

Disabling Status Bar Display

By default, IE4+ shows a status bar with varying amounts of information. If this poses a security breach (by showing information an administrator might wish kept hidden, such as a URL), you can disable the status bar display on a per-user basis by tweaking the Registry as follows:

1. Launch Regedt32.

2. Select the Window menu option for **HKEY_CURRENT_USER**.

3. Use the tree control in the left-hand window to navigate to the **SOFTWARE\Microsoft\Internet Explorer\Main** subkey. Click on the subkey to select it and display its values in the right-hand window.

4. Locate the **Show_StatusBar** value entry. Use the String Editor to change the Show_StatusBar value to No to disable the status bar display for the user.

Disabling Toolbar Display

By default, IE4+ shows a toolbar with varying amounts of program control. If this poses a security breach (for example, if it allows the user to view source code for Web pages), you can disable the toolbar display on a per-user basis by tweaking the Registry as follows:

1. Launch Regedt32.

2. Select the Window menu option for **HKEY_CURRENT_USER**.

3. Use the tree control in the left-hand window to navigate to the **SOFTWARE\Microsoft\Internet Explorer\Main** subkey. Click on the subkey to select it and display its values in the right-hand window.

4. Locate the **Show_Toolbar** value entry. Use the String Editor to change the Show_Toolbar value to No to disable the toolbar for a user.

Disabling Status Bar URL Display

By default, IE4+ shows the URL of a page being downloaded in the status bar. If this poses a security breach (that is, if the URL contains sensitive information), you can disable the URL display on a per-user basis by tweaking the Registry as follows:

1. Launch Regedt32.

2. Select the Window menu option for **HKEY_CURRENT_USER**.

3. Use the tree control in the left-hand window to navigate to the **SOFTWARE\Microsoft\Internet Explorer\Main** subkey. Click on the subkey to select it and display its values in the right-hand window.

4. Locate the **Show_URLinStatusBar** value entry. Use the String Editor to change the Show_URLinStatusBar value to No to disable URL display during downloading.

WARNING! Adjusting the Show_URLinStatusBar feature does not affect the display of the URL Address bar.

18. Internet Explorer 4+

Disabling Address Bar Display

By default, IE4+ shows the URL of the page after it is downloaded in the Address bar. If this poses a security breach (that is, if the URL contains sensitive information, such as a hidden directory), you can disable the URL display on a per-user basis by tweaking the Registry as follows:

1. Launch Regedt32.

2. Select the Window menu option for **HKEY_CURRENT_USER**.

3. Use the tree control in the left-hand window to navigate to the **SOFTWARE\Microsoft\Internet Explorer\Main** subkey. Click on the subkey to select it and display its values in the right-hand window.

4. Locate the **Show_URLToolBar** value entry. Use the String Editor to change the Show_URLToolBar value to No to disable the display of the Address bar.

WARNING! *Disabling the Show_URLToolBar feature does not suppress the display of URLs in the status bar during downloading.*

Setting the Start Page

IE4+ comes configured to display a given URL (its *start page*) when users open IE. To change the default start page, you can change the Registry as follows:

1. Launch Regedt32.

2. Select the Window menu option for **HKEY_CURRENT_USER**.

3. Use the tree control in the left-hand window to navigate to the **SOFTWARE\Microsoft\Internet Explorer\Main** subkey. Click on the subkey to select it and display its values in the right-hand window.

4. Locate the **Start Page** value entry. Use the String Editor to change the Start Page value to the desired URL for the user and/or company.

TIP: *A start page URL can be a local HTML page.*

Configuring Active Desktop Wallpaper

IE4+'s Active Desktop introduces the capability to use HTML, JPEG, and GIF files as desktop wallpaper. You can customize this feature on a per-user basis by tweaking the Registry as follows:

1. Launch Regedt32.

2. Select the Window menu option for **HKEY_CURRENT_USER**.

3. Use the tree control in the left-hand window to navigate to the **SOFTWARE\Microsoft\Internet Explorer\Desktop\General** subkey. Click on the subkey to select it and display its values in the right-hand window.

4. Locate the **Wallpaper** value entry. Use the String Editor to change the Wallpaper value to the full path and file name of the HTML, JPG, or GIF file you want to use as the Active Desktop wallpaper.

Setting Active Desktop Wallpaper Tiling

As with BMP-based wallpaper, Active Desktop wallpaper files can be tiled (displayed multiple times to cover the entire desktop). You can set the tiling feature on a per-user basis by tweaking the Registry as follows:

1. Launch Regedt32.

2. Select the Window menu option for **HKEY_CURRENT_USER**.

3. Use the tree control in the left-hand window to navigate to the **SOFTWARE\Microsoft\Internet Explorer\Desktop\General** subkey. Click on the subkey to select it and display its values in the right-hand window.

4. Locate the **TileWallpaper** value entry. Use the String Editor to change the TileWallpaper value to1 to turn on wallpaper tiling, or 0 to turn it off.

Configuring Hard Copy Page Settings

IE4 uses the Registry to store margin, footer, and header settings for hard-copy printouts of pages displayed in IE. You can configure the Registry's settings to meet the specialized needs of a user (for example, the legal department might need to include a disclaimer on each page of a printout). To do so, follow these steps:

1. Launch Regedt32.

2. Select the Window menu option for **HKEY_CURRENT_USER**.

3. Use the tree control in the left-hand window to navigate to the **SOFTWARE\Microsoft\Internet Explorer\PageSetup** subkey. Click on the subkey to select it and display its values in the right-hand window.

4. Locate the **footer** value entry. Use the String Editor to change the footer value to the desired setting (which can include literal text as well as wildcards for the URL and other elements).

5. Locate the **header** value entry. Use the String Editor to change the header value to the desired setting (which can include literal text as well as wildcards for the URL and other elements).

6. Locate the **margin_bottom** value entry. Use the String Editor to change the margin_bottom value to the desired setting (in decimal fractions of an inch).

7. Locate the **margin_top** value entry. Use the String Editor to change the margin_top value to the desired setting (in decimal fractions of an inch).

8. Locate the **margin_left** value entry. Use the String Editor to change the margin_left value to the desired setting (in decimal fractions of an inch).

9. Locate the **margin_right** value entry. Use the String Editor to change the margin_right value to the desired setting (in decimal fractions of an inch).

Setting Safety Warning Level

Some users of IE4+ prefer to use their browsers at a riskier safety level than others, allowing unsigned active content on their workstations without warning or rejection. You can override this risky behavior via the Registry as follows:

1. Launch Regedt32.

2. Select the Window menu option for **HKEY_CURRENT_USER**.

3. Use the tree control in the left-hand window to navigate to the **SOFTWARE\Microsoft\Internet Explorer\Security** subkey. Click on the subkey to select it and display its values in the right-hand window.

4. Locate the **Safety Warning Level** value entry. Use the String Editor to change the Safety Warning Level value to Query to turn on security so that any unsafe content generates a warning to the user.

Configuring Display Colors

Although many Web pages have specific custom color schemes, others do not. Users can specify default color settings for Web pages viewed within their IE4+ browser. You can use the Registry to specify the default colors. To do so, follow these steps:

1. Launch Regedt32.

2. Select the Window menu option for **HKEY_CURRENT_USER**.

3. Use the tree control in the left-hand window to navigate to the **SOFTWARE\Microsoft\Internet Explorer\Settings** subkey. Click on the subkey to select it and display its values in the right-hand window.

4. Locate the value entry for the color setting you want to change. Table 18.1 shows the IE4 display color values. Use the String Editor to change each value to an RGB (red, green, blue) color, which consists of three comma-delimited color values.

Table 18.1   Value entries for IE4+ display colors.

Value	Display Element
Anchor color	Unvisited links
Anchor color visited	Links that have been clicked
Background color	Page background
Text color	Text

18. Internet Explorer 4+

Setting URL AutoComplete Candidates

One of the more useful features of IE4+ is *URL AutoComplete*. AutoComplete uses a list of recently typed URLs to assist users when they type URLs in the Address bar. A neat trick for administrators is to preset a group of often-typed URLs prior to a user typing them in. Here's how to add preset URLs to the Registry for the AutoComplete feature to refer to:

1. Launch Regedt32.

2. Select the Window menu option for **HKEY_CURRENT_USER**.

3. Use the tree control in the left-hand window to navigate to the **SOFTWARE\Microsoft\Internet Explorer\TypedURLs** subkey. Click on the subkey to select it and display its values in the right-hand window.

4. Locate the **url1** value entry. Use the String Editor to change the url1 value to one of the preset URLs you would like to supply for users.

TIP: You can add url# entries to the Registry if you want to add multiple URL addresses for the AutoComplete feature.

Setting the File Cache Directory

Each IE4+ user can have a private set of directories to store various files related to the browser's functionality. You can specify the folder used to store downloaded (cached) Internet files, such as Web pages, by following these steps:

1. Launch Regedt32.

2. Select the Window menu option for **HKEY_CURRENT_USER**.

3. Use the tree control in the left-hand window to navigate to the **SOFTWARE\Microsoft\Windows\CurrentVersion\Explorer\User Shell Folders** subkey. Click on the subkey to select it and display its values in the right-hand window.

4. Locate the **Cache** value entry. Use the String Editor to change the Cache value to the desired folder for the user.

TIP: You can modify the %USERPROFILE% environment variable to change the root path of the folder in the Cache entry.

18. Internet Explorer 4+

Setting the Cookies Directory

You can specify the folder used to store the special text strings used to persist Web page settings (cookies) by following these steps:

1. Launch Regedt32.

2. Select the Window menu option for **HKEY_CURRENT_USER**.

3. Use the tree control in the left-hand window to navigate to the **SOFTWARE\Microsoft\Windows\CurrentVersion\Explorer\User Shell Folders** subkey. Click on the subkey to select it and display its values in the right-hand window.

4. Locate the **Cookies** value entry. Use the String Editor to change the Cookies value to specify the desired folder for the user.

TIP: You can modify the %USERPROFILE% environment variable to change the root path of the folder in the Cookies entry.

Setting the Favorites Directory

You can specify the folder used to store the list of URLs a user wants to revisit (Favorites) by following these steps:

1. Launch Regedt32.

2. Select the Window menu option for **HKEY_CURRENT_USER**.

3. Use the tree control in the left-hand window to navigate to the **SOFTWARE\Microsoft\Windows\CurrentVersion\Explorer\User Shell Folders** subkey. Click on the subkey to select it and display its values in the right-hand window.

4. Locate the **Favorites** value entry. Use the String Editor to change the Favorites value to the desired folder for that user. Remember that multiple menus in the Favorites system end up becoming separate folders beneath the root Favorites directory.

TIP: You can modify the %USERPROFILE% environment variable to change the root path of the folder in the Favorites entry.

---

18. Internet Explorer 4+

Setting the History Directory

You can specify the folder used to store previously visited URLs as Internet shortcuts (History) by following these steps:

1. Launch Regedt32.

2. Select the Window menu option for **HKEY_CURRENT_USER**.

3. Use the tree control in the left-hand window to navigate to the **SOFTWARE\Microsoft\Windows\CurrentVersion\Explorer\User Shell Folders** subkey. Click on the subkey to select it and display its values in the right-hand window.

4. Locate the **History** value entry. Use the String Editor to change the History value to the desired folder for the user.

TIP: You can modify the %USERPROFILE% environment variable to change the root path of the folder in the History entry.

Setting the Active Desktop Start Menu Startup Directory

You can specify the folder used to store shortcuts to the programs and/or documents to be started each time a user logs on. To do so, follow these steps:

1. Launch Regedt32.

2. Select the Window menu option for **HKEY_CURRENT_USER**.

3. Use the tree control in the left-hand window to navigate to the **SOFTWARE\Microsoft\Windows\CurrentVersion\Explorer\User Shell Folders** subkey. Click on the subkey to select it and display its values in the right-hand window.

4. Locate the **Startup** value entry. Use the String Editor to change the Startup value to the desired folder for the user.

TIP: You can modify the %USERPROFILE% environment variable to change the root path of the folder in the Startup entry.

Setting the Active Desktop Start Menu Directory

You can specify the folder used to store the shortcuts and files that are included on the Start menu for a user. To do so, follow these steps:

1. Launch Regedt32.

2. Select the Window menu option for **HKEY_CURRENT_USER**.

3. Use the tree control in the left-hand window to navigate to the **SOFTWARE\Microsoft\Windows\CurrentVersion\Explorer\User Shell Folders** subkey. Click on the subkey to select it and display its values in the right-hand window.

4. Locate the **StartMenu** value entry. Use the String Editor to change the StartMenu value to the desired folder for the user.

TIP: You can modify the %USERPROFILE% environment variable to change the root path of the folder in the StartMenu entry.

Setting the Active Desktop Send to Directory

You can specify the folder used to store applications and folders to appear on the Send To menu option for various Active Desktop applications (such as IE4+). To do so, follow these steps:

1. Launch Regedt32.

2. Select the Window menu option for **HKEY_CURRENT_USER**.

3. Use the tree control in the left-hand window to navigate to the **SOFTWARE\Microsoft\Windows\CurrentVersion\Explorer\User Shell Folders** subkey. Click on the subkey to select it and display its values in the right-hand window.

4. Locate the **SendTo** value entry. Use the String Editor to change the SendTo value to the desired folder for the user.

TIP: You can modify the %USERPROFILE% environment variable to change the root path of the folder in the SendTo entry. Also, if the value does not exist, it can be created.

18. Internet Explorer 4+

Setting the Active Desktop Recent Files Directory

You can specify the folder used to store a list of shortcuts to recently opened document files by following these steps:

1. Launch Regedt32.
2. Select the Window menu option for **HKEY_CURRENT_USER**.
3. Use the tree control in the left-hand window to navigate to the **SOFTWARE\Microsoft\Windows\CurrentVersion\Explorer\User Shell Folders** subkey. Click on the subkey to select it and display its values in the right-hand window.
4. Locate the **Recent** value entry. Use the String Editor to change the Recent value to the desired folder for the user.

TIP: You can modify the %USERPROFILE% environment variable to change the root path of the folder in the Recent entry. Also, if the value does not exist, it can be created.

Setting the Active Desktop Programs Directory

You can specify the folder used to store the shortcuts and folders to be shown in the Programs menu of the Start menu. To do so, follow these steps:

1. Launch Regedt32.
2. Select the Window menu option for **HKEY_CURRENT_USER**.
3. Use the tree control in the left-hand window to navigate to the **SOFTWARE\Microsoft\Windows\CurrentVersion\Explorer\User Shell Folders** subkey. Click on the subkey to select it and display its values in the right-hand window.
4. Locate the **Programs** value entry. Use the String Editor to change the Programs value to the desired folder for the user.

TIP: You can modify the %USERPROFILE% environment variable to change the root path of the folder in the Programs entry.

---

18. Internet Explorer 4+

Setting the Active Desktop **PrintHood** Directory

You can specify the folder used to store the list of printers available to the current workstation and user (**PrintHood**) by following these steps:

1. Launch Regedt32.

2. Select the Window menu option for **HKEY_CURRENT_USER**.

3. Use the tree control in the left-hand window to navigate to the **SOFTWARE\Microsoft\Windows\CurrentVersion\Explorer\User Shell Folders** subkey. Click on the subkey to select it and display its values in the right-hand window.

4. Locate the **PrintHood** value entry. Use the String Editor to change the PrintHood value to the desired folder for the user.

TIP: You can modify the %USERPROFILE% environment variable to change the root path of the folder in the PrintHood entry.

Setting the Active Desktop Personal Directory

You can specify the folder used to store personal folders (such as My Documents) by following these steps:

1. Launch Regedt32.

2. Select the Window menu option for **HKEY_CURRENT_USER**.

3. Use the tree control in the left-hand window to navigate to the **SOFTWARE\Microsoft\Windows\CurrentVersion\Explorer\User Shell Folders** subkey. Click on the subkey to select it and display its values in the right-hand window.

4. Locate the **Personal** value entry. Use the String Editor to change the Personal value to the desired folder for the user.

TIP: You can modify the %USERPROFILE% environment variable to change the root path of the folder in the Personal entry.

Setting the Active Desktop **NetHood** Directory

You can specify the folder used to store the list of available computers (**NetHood**) for the current user and workstation. To do so, follow these steps:

1. Launch Regedt32.

2. Select the Window menu option for **HKEY_CURRENT_USER**.

3. Use the tree control in the left-hand window to navigate to the **SOFTWARE\Microsoft\Windows\CurrentVersion\Explorer\User Shell Folders** subkey. Click on the subkey to select it and display its values in the right-hand window.

4. Locate the **NetHood** value entry. Use the String Editor to change the NetHood value to the desired folder for the user.

TIP: *You can modify the %USERPROFILE% environment variable to change the root path of the folder in the NetHood entry.*

Chapter 19

Microsoft Transaction Server

In Brief

Microsoft Transaction Server (MTS) has evolved from being an add-on for Windows NT to a core part of the Windows 2000 operating system. MTS allows COM (Component Object Model) components to participate in special database manipulations called *transactions*, which have the vital capability of transparently restoring their previous state if a problem occurs during any operation of the transaction. MTS uses an Explorer-like user interface, but stores most of its information in the Registry. MTS is made up of packages (groups of components that are COM servers). It features a sophisticated new security feature called *roles* and allows for remote installation of MTS package components.

MTS Packages

An MTS package is a group of one or more MTS components that participate in a transaction. Components are COM servers specially programmed to interact with MTS. Packages are created using the MTS Explorer's user interface. MTS "intercepts" calls made to the COM runtime system (actually replacing a Registry entry with its own) for components in a package so that MTS does the work of creating their instance in memory rather than letting COM do it. Packages have the following elements:

- *Name*—Used for administration only.

- *GUID (Globally Unique Identifier)*—Special hexadecimal number string that represents the system name. The GUID values in the Registry are surrounded by curly brackets; for example, {976909E1-650D-11d3-B7EF-00E02916C424}. This is called Registry Format. The GUID you obtain from the developer might not contain these characters.

TIP: *You can obtain the GUID value of a package from the MTS administrator user interface by selecting a package and displaying its Properties dialog box. The General tab identifies a Package ID, which is the GUID you need.*

- *Associated components*—The COM servers in the package, listed by GUID.

- *Associated roles*—Names of the security roles used by the MTS security system.

- *Administration settings*—For example, change and delete permissions.
- *System settings*—For example, privileges and memory management.
- *COM settings*—For example, transaction behavior and threading models.

Security Roles

Unlike Windows NT/2000, which uses either username/password combinations or system SID (Security Identifier) values, MTS uses text-based security called roles. Roles are stored in the Registry as a combination of GUID, text names, usernames, and SID values. When a component needs to verify that its user's account is one that should have a given level of access, the program code makes an API function call to ask MTS whether the current user is listed as part of the role. If the user is part of the role, the application can continue; if not, the application can take appropriate error measures.

Remote Installation

You can install and access MTS components remotely. In either case, special Registry entries are added to allow MTS to record where the actual binary code lives. The MTS Explorer can export special files that automatically configure MTS on the remote machine so the Explorer can connect with the MTS installation at runtime.

MTS Components

The heart of MTS is the collection of components that use it. These components are COM software elements and have the following unique characteristics:

- *Component DLL*—All COM components for MTS must have a DLL (in-process) server provided for them. The server name and path are stored in the Registry, both for COM and MTS.
- *Component interfaces*—Each MTS component supports one or more COM interfaces that COM uses to provide software services to its client applications. Each interface has a unique GUID value (a string of hexadecimal digits) that is guaranteed to be unique across any combination of machines.
- *Component method*—Each MTS COM component interface supports a group of functions called *methods*. These functions have normal text names and clients of MTS components use them to provide the desired MTS-oriented services.

Immediate Solutions

Locating MTS Installation Files

One Registry setting that can be particularly useful to administrators specifies the location at which MTS files were installed during setup. To access this Registry setting, follow these steps:

1. Launch Regedt32.

2. Select the Window menu option for **HKEY_LOCAL_MACHINE**.

3. Use the tree control in the left-hand window to navigate to the **SOFTWARE\Microsoft\Transaction Server\Setup** subkey. Click on the subkey to select it and display its values in the right-hand window.

4. Locate the **Install Path** value entry. Verify that the MTS files are indeed where MTS expects them to be and, if necessary, reset the entry using the String Editor.

Locating MTS Source Files

Another Registry setting specifies where MTS and Windows 2000 expect the source files to be found (for upgrades or re-installations). To access this Registry setting, follow these steps:

1. Launch Regedt32.

2. Select the Window menu option for **HKEY_LOCAL_MACHINE**.

3. Use the tree control in the left-hand window to navigate to the **SOFTWARE\Microsoft\Transaction Server\Setup** subkey. Click on the subkey to select it and display its values in the right-hand window.

4. Locate the **Source Path** value entry. The Source Path value is set during installation. If the source file(s) are now in a different location, you can change the entry using the String Editor.

Determining Installed MTS Version Numbers

Another Registry setting specifies the version number of MTS (which, if corrupted, can cause problems for the system). To access this Registry setting, follow these steps:

1. Launch Regedt32.

2. Select the Window menu option for **HKEY_LOCAL_MACHINE**.

3. Use the tree control in the left-hand window to navigate to the **SOFTWARE\Microsoft\Transaction Server\Setup\Product Version** subkey. Click on the subkey to select it and display its values in the right-hand window.

4. Locate the **Major** value entry or the **Minor** value entry and compare it with the expected information.

TIP: *The Major and Minor values are in hexadecimal, with version 1 being 0x1000 and version 2 being 0x2000. The Minor version number appears to be the build number for the current installation and is not currently used.*

Determining if an MTS Component Has Been Installed

Another Registry setting lists the components currently registered with the MTS system. To access this Registry setting, follow these steps:

1. Launch Regedt32.

2. Select the Window menu option for **HKEY_LOCAL_MACHINE**.

3. Use the tree control in the left-hand window to navigate to the **SOFTWARE\Microsoft\Transaction Server\Components** subkey. Double-click on the subkey to expand it and display its values in the right-hand window.

4. Obtain the **GUID** value (a unique hexadecimal number string) of the component in question. Check the list of subkeys for components until you find one that matches the value you selected. If you find a match, the component is currently installed; if you do not find a match, the component has been lost and must be reinstalled using the MTS administrator user interface.

Locating the Type Library for an Installed MTS Component

Another Registry setting specifies the location of the type library file for an installed MTS component (needed for MTS to be accessed by Visual Basic and Java applications). To find the path to the type library file for an MTS component, follow these steps:

1. Launch Regedt32.

2. Select the Window menu option for **HKEY_LOCAL_MACHINE**.

3. Use the tree control in the left-hand window to navigate to the **SOFTWARE\Microsoft\Transaction Server\Components** subkey. Double-click on the subkey to expand it and display its values in the right-hand window.

4. Obtain the **GUID** value (a unique hexadecimal number string) of the component in question. Check the list of subkeys for components until you find one that matches the value you selected. If you find a match, the component is currently installed; if you do not find a match, the component has been lost and you must reinstall it using the MTS administrator user interface.

5. Click on the subkey selected in Step 4 to display its values in the right-hand window. Locate the **Typelib** value entry and be sure it is the right one for that component. If not, use the String Editor to change the path (and file name, if necessary) to the appropriate value(s).

Determining if an MTS Component's Transaction Attribute Is Set Correctly

Another Registry setting holds the transaction requirements of the component (set via the MTS administration user interface). To view a component's transaction requirements, follow these steps:

1. Launch Regedt32.

2. Select the Window menu option for **HKEY_LOCAL_MACHINE**.

3. Use the tree control in the left-hand window to navigate to the **SOFTWARE\Microsoft\Transaction Server\Components** subkey. Double-click on the subkey to expand it and display its values in the right-hand window.

4. Obtain the **GUID** value of the component in question. Check the list of subkeys for components until you find one that matches the value you selected. If you find a match, the component is currently installed; if you do not find a match, the component has been lost and you must reinstall it using the MTS administrator user interface.

5. Click on the subkey selected in Step 4 to display its values in the right-hand window. Locate the **Transaction** value entry and be sure it is the right one for the selected component. If the Transaction setting is incorrect, use the String Editor to change the setting to the required value. Table 19.1 shows possible Transaction values along with MTS administration user interface equivalents.

Table 19.1   MTS Transaction Registry entry values.

Value	User Interface Equivalent
Not Supported	The component does not support Transactions.
Requires New	The component requires a new Transaction each time it is activated.
Supported	The component supports Transactions, but does not require one.
Required	The component requires a Transaction, but does not require a new one.

Determining an MTS Component's Threading Model

Another Registry setting specifies the threading model a component supports. The threading model can have an immense impact on performance. This is because components that can be used in a multithreaded environment without causing problems are able to execute simultaneously. To view the threading model supported by a component, follow these steps:

1. Launch Regedt32.

2. Select the Window menu option for **HKEY_LOCAL_MACHINE**.

3. Use the tree control in the left-hand window to navigate to the **SOFTWARE\Microsoft\Transaction Server\Components** subkey. Double-click on the subkey to expand it and display its values in the right-hand window.

4. Obtain the **GUID** value (a unique hexadecimal number string) of the component in question. Check the list of subkeys for

components until you find one that matches the value you selected. If you find a match, the component is currently installed; if you do not find a match, the component has been lost and you must reinstall it using the MTS administrator user interface.

5. Click on the subkey selected in Step 4 to display its values in the right-hand window. Locate the **ThreadingModel** value entry and be sure it is the right one for that component (as provided by the documentation or developer). If the ThreadingModel value is incorrect, use the String Editor to change it to the appropriate value. Table 19.2 shows possible ThreadingModel values along with a description of each value.

Table 19.2   MTS ThreadingModel Registry entry values.

Value	Description
Single	The component does not support threading.
Apartment	The component supports threading.
Both	The component supports threading, but is not degraded without it.
Free	The component requires threading to operate.

Obtaining an MTS Component's Programmatic Identification String

Another Registry setting specifies the ProgID (programmatic identification string) used by Visual Basic applications. If you are a developer, you certainly are interested in retrieving this information. To access this Registry setting, follow these steps:

1. Launch Regedt32.

2. Select the Window menu option for **HKEY_LOCAL_MACHINE**.

3. Use the tree control in the left-hand window to navigate to the **SOFTWARE\Microsoft\Transaction Server\Components** subkey. Double-click on the subkey to expand it and display its values in the right-hand window.

4. Obtain the **GUID** value (a unique hexadecimal number string) of the component in question. Check the list of subkeys for components until you find one that matches the value you selected. If you find a match, the component is currently installed; if you do

not find a match, the component has been lost and you must reinstall it using the MTS administrator user interface.

5. Click on the subkey selected in Step 4 to display its values in the right-hand window. Locate the **ProgID** value entry and be sure it is the right one for that component (as provided by the documentation or developer). If the ProgID value is incorrect, use the String Editor to change it to the appropriate value.

Locating the Folder for Installed MTS Packages

Another Registry setting specifies the location of the directory MTS expects to use for installed packages. To view this directory's path name, follow these steps:

1. Launch Regedt32.

2. Select the Window menu option for **HKEY_LOCAL_MACHINE**.

3. Use the tree control in the left-hand window to navigate to the **SOFTWARE\Microsoft\Transaction Server\Local Computer*[computer name]*** subkey, where *[computer name]* is the name of the local computer. Double-click on the subkey to expand it and display its values in the right-hand window.

4. Locate the **Package Directory** value entry. If the Package Directory value does not specify the directory the packages are actually using, use the String Editor to change the value.

Locating the Folder for Remote MTS Components

Another Registry setting specifies the location of the directory where MTS expects to keep information about its remote components. To access this Registry setting, follow these steps:

1. Launch Regedt32.

2. Select the Window menu option for **HKEY_LOCAL_MACHINE**.

3. Use the tree control in the left-hand window to navigate to the **SOFTWARE\Microsoft\Transaction Server\Local Computer\\[computer name]** subkey, where *[computer name]* is the name of the local computer. Double-click on the subkey to expand it and display its values in the right-hand window.

4. Locate the **Remote Components** value entry. If the Remote Components value does not specify the directory the remote components are actually using, use the String Editor to change the value.

Determining the Oracle DLLs Used by MTS

Another Registry setting lists the DLL names for Oracle functionality that MTS is using. (Oracle is one of the two major relational database management systems that support MTS natively.) If these settings are the wrong ones, errors happen without apparent cause. To access this Registry setting, follow these steps:

1. Launch Regedt32.

2. Select the Window menu option for **HKEY_LOCAL_MACHINE**.

3. Use the tree control in the left-hand window to navigate to the **SOFTWARE\Microsoft\Transaction Server\Local Computer\\[computer name]** subkey, where *[computer name]* is the name of the local computer. Double-click on the subkey to expand it and display its values in the right-hand window.

4. Locate the **OracleSqlLib** value entry and the **OracleXaLib** value entry. Make sure the values are the proper ones for MTS (as given in the MTS documentation). If not, determine whether the DLLs named in the values are on the system by using Explorer to examine the given path location. If the values are on the system, use the String Editor to change the entries to the proper values; otherwise, reinstall Oracle and then reinstall MTS.

Determining the User ID for an MTS Package

Another Registry setting is the **UserID** value for a given server in an MTS application (that is, the account under which it will run unless using impersonation or some other SID-altering arrangement):

1. Launch Regedt32.

2. Select the Window menu option for **HKEY_LOCAL_MACHINE**.

3. Use the tree control in the left-hand window to navigate to the **SOFTWARE\Microsoft\Transaction Server\Packages** subkey. Double-click on the subkey to expand it and display its values in the right-hand window.

4. Obtain the **GUID** value (a unique hexadecimal number string) of the package in question. Check the list of subkeys for packages until you find one that matches the value you selected. If you find a match, the package is currently installed; if you do not find a match, the component has been lost and you must reinstall it using the MTS administrator user interface.

5. Click on the subkey you selected in Step 4 to display its values in the right-hand window. Locate the **UserID** value entry and be sure it is the right one for the selected package (as provided by the documentation or developer). If the UserID value is incorrect, use the String Editor to change it to the appropriate value.

TIP: You can obtain the GUID value of a given package from the MTS administrator user interface by selecting a package of interest and displaying its Properties dialog box. The General tab identifies a Package ID, which is the GUID you need.

Determining whether an MTS Package Is a System Package

Another Registry setting indicates whether an MTS package has been made a system package. If it has, this can have a substantial impact on the package's behavior because system packages run under the system account and can therefore bypass a number of security protections. To see whether an MTS package is a system package, follow these steps:

1. Launch Regedt32.

2. Select the Window menu option for **HKEY_LOCAL_MACHINE**.

3. Use the tree control in the left-hand window to navigate to the **SOFTWARE\Microsoft\Transaction Server\Packages** subkey. Double-click on the subkey to expand it and display its values in the right-hand window.

4. Obtain the **GUID** value (a unique hexadecimal number string) of the package in question. Check the list of subkeys for packages until you find one that matches. If you find a match, the package is currently installed; if you do not find a match, the package has been lost and you must reinstall it using the MTS administrator user interface.

5. Click on the subkey selected in Step 4 to display its values in the right-hand window. Locate the **System** value entry and be sure it is the right one for that package (as provided by the documentation or developer). If the System value is incorrect, use the String Editor to change it to the appropriate value (Y or N).

TIP: You can obtain the GUID value of a package from the MTS administrator user interface by selecting a package and displaying its Properties dialog box. The General tab identifies a Package ID, which is the GUID you need.

Determining whether Security Is Enabled for an MTS Package

Another Registry setting indicates whether security is enabled for a given MTS package. To see if security is enabled for an MTS package, follow these steps:

1. Launch Regedt32.

2. Select the Window menu option for **HKEY_LOCAL_MACHINE**.

3. Use the tree control in the left-hand window to navigate to the **SOFTWARE\Microsoft\Transaction Server\Packages** subkey. Double-click on the subkey to expand it and display its values in the right-hand window.

4. Obtain the **GUID** value (a unique hexadecimal number string) of the package in question. Check the list of subkeys for packages until you find one that matches the value you selected. If you find a match, the package is currently installed; if you do not find a match, the package has been lost and you must reinstall it using the MTS administrator user interface.

5. Click on the subkey selected in Step 4 to display its values in the right-hand window. Locate the **SecurityEnabled** value

entry and be sure it is the right one for that package (as provided by the documentation or developer). If the SecurityEnabled value is incorrect, use the String Editor to change it to the appropriate value (Y or N).

TIP: You can obtain the GUID value of a package from the MTS administrator user interface by selecting a package and displaying its Properties dialog box. The General tab identifies a Package ID, which is the GUID you need.

Enabling MTS to Shut Down an MTS Package When It Is No Longer Needed

Another Registry setting specifies whether MTS can remove a package's components from memory when they are no longer needed. To access this Registry setting, follow these steps:

1. Launch Regedt32.

2. Select the Window menu option for **HKEY_LOCAL_MACHINE**.

3. Use the tree control in the left-hand window to navigate to the **SOFTWARE\Microsoft\Transaction Server\Packages** subkey. Double-click on the subkey to expand it and display its values in the right-hand window.

4. Obtain the **GUID** value (a unique hexadecimal number string) of the package in question. Check the list of subkeys for packages until you find one that matches the value you selected. If you find a match, the package is currently installed; if you do not find a match, the package has been lost and you must reinstall it using the MTS administrator user interface.

5. Click on the subkey selected in Step 4 to display its values in the right-hand window. Locate the **NeverShutdown** value entry and be sure it is the right one for that package (as provided by the documentation or developer). If the NeverShutdown value is incorrect, use the String Editor to change it to the appropriate value (Y or N).

TIP: You can obtain the GUID value of a package from the MTS administrator user interface by selecting a package and displaying its Properties dialog box. The General tab identifies a Package ID, which is the GUID you need.

Disabling Deletion of an MTS Package

Another Registry setting controls whether anyone can delete a package from the MTS administrator user interface. To access this Registry setting, follow these steps:

1. Launch Regedt32.

2. Select the Window menu option for **HKEY_LOCAL_MACHINE**.

3. Use the tree control in the left-hand window to navigate to the **SOFTWARE\Microsoft\Transaction Server\Packages** subkey. Double-click on the subkey to expand it and display its values in the right-hand window.

4. Obtain the **GUID** value (a unique hexadecimal number string) of the package in question. Check the list of subkeys for packages until you find one that matches the value you selected. If you find a match, the package is currently installed; if you do not find a match, the package has been lost and you must reinstall it using the MTS administrator user interface.

5. Click on the subkey selected in Step 4 to display its values in the right-hand window. Locate the **Deleteable** value entry and be sure it is the right one for that package (as provided by the documentation or developer). If the Deleteable value is incorrect, use the String Editor to change it to the appropriate value (Y or N).

TIP: You can obtain the GUID value of a package from the MTS administrator user interface by selecting a package and displaying its Properties dialog box. The General tab identifies a Package ID, which is the GUID you need.

WARNING! If you make a component deleteable, it can be deleted either from the user interface or by administration scripts. If you make a component nondeleteable, it can be deleted only after the Registry setting is changed (directly or via the user interface with administrator privileges).

Disabling Changing of an MTS Package

Another Registry setting indicates whether the elements of a package can be changed. To access this Registry setting, follow these steps:

1. Launch Regedt32.

2. Select the Window menu option for **HKEY_LOCAL_MACHINE**.

3. Use the tree control in the left-hand window to navigate to the **SOFTWARE\Microsoft\Transaction Server\Packages** subkey. Double-click on the subkey to expand it and display its values in the right-hand window.

4. Obtain the **GUID** value (a unique hexadecimal number string) of the package in question. Check the list of subkeys for packages until you find one that matches the value you selected. If you find a match, the package is currently installed; if you do not find a match, the package has been lost and you must reinstall it using the MTS administrator user interface.

5. Click on the subkey selected in Step 4 to display its values in the right-hand window. Locate the **Changeable** value entry and be sure it is the right one for that package (as provided by the documentation or developer). If the Changeable value is incorrect, use the String Editor to change it to the appropriate value (Y or N).

TIP: You can obtain the GUID value of a package from the MTS administrator user interface by selecting a package and displaying its Properties dialog box. The General tab identifies a Package ID, which is the GUID you need.

WARNING! If you make a component changeable, it can be changed either from the user interface or by administration scripts. If you make a component nonchangeable, it can be altered only after the Registry setting is changed (directly or via the user interface with administrator privileges).

Determining the Names of MTS Component Roles

Another Registry setting assists in determining the text names assigned to MTS security roles for a package. These values must exactly match those expected in the program's internal code, or the program will not detect that a given user is in such a role and will refuse to operate properly. To access this Registry setting, follow these steps:

1. Launch Regedt32.

2. Select the Window menu option for **HKEY_LOCAL_MACHINE**.

3. Use the tree control in the left-hand window to navigate to the **SOFTWARE\Microsoft\Transaction Server\Packages** subkey. Double-click on the subkey to expand it and display its values in the right-hand window.

4. Obtain the **GUID** value (a unique hexadecimal number string) of the package in question. Check the list of subkeys for packages until you find one that matches the selected value. If you find a match, the package is currently installed; if you do not find a match, the package has been lost and you must reinstall it using the MTS administrator user interface. Repeat this process on each of the selected package's subkeys, checking for the role's GUID value (obtained via the Properties dialog for the role in the administrator user interface as the Role ID); if you don't find this GUID entry, you must use the MTS Explorer to re-create the role entry for that component.

5. Click on the second subkey selected in Step 4 to display its values in the right-hand window. Locate the **Name** value entry and be sure it is the right one for the Role ID (as provided by the documentation or developer). If the Name value is incorrect, use the String Editor to change it to the appropriate value.

TIP: *You can obtain the GUID value of a package from the MTS administrator user interface by selecting a package and displaying its Properties dialog box. The General tab identifies a Package ID, which is the GUID you need.*

Determining the Users Assigned to an MTS Role

Another Registry setting specifies which Windows 2000 usernames are considered members of a given MTS security role (determined automatically by MTS at runtime from the application's SID token). To view the Windows 2000 usernames that belong to an MTS security role, follow these steps:

1. Launch Regedt32.

2. Select the Window menu option for **HKEY_LOCAL_MACHINE**.

3. Use the tree control in the left-hand window to navigate to the **SOFTWARE\Microsoft\Transaction Server\Packages** subkey. Double-click on the subkey to expand it and display its values in the right-hand window.

4. Obtain the **GUID** value (a unique hexadecimal number string) of the package in question. Check the list of subkeys for packages until you find one that matches the value you selected. If you find a match, the package is currently installed; if you do not find a match, the package has been lost and you must reinstall it using the MTS administrator user interface. Repeat this process on each of the selected package's subkeys, checking for the role's GUID value (obtained via the Properties dialog box for the role in the administrator user interface as the Role ID). If you don't find this GUID entry, you must use the MTS Explorer to re-create the role entry for that component.

5. Click on the second subkey selected in Step 4 to display its values in the right-hand window; then, open its **Users** subkey. A list of keys that are the usernames of the users assigned to the role should appear. You can add usernames using the MTS administrator user interface.

TIP: *You can obtain the GUID value of a package from the MTS administrator user interface by selecting a package and displaying its Properties dialog. The General tab identifies a Package ID, which is the GUID you need.*

Determining if a COM Server Has Been Assigned to an MTS Package

Another Registry setting contains the GUID value for a COM server that supports MTS and indicates whether it is part of a given MTS package. To access this Registry value, follow these steps:

1. Launch Regedt32.

2. Select the Window menu option for **HKEY_LOCAL_MACHINE**.

3. Use the tree control in the left-hand window to navigate to the **SOFTWARE\Microsoft\Transaction Server\Packages** subkey. Double-click on the subkey to expand it and display its values in the right-hand window.

4. Obtain the **GUID** value (a unique hexadecimal number string) of the package in question. Check the list of subkeys for packages until you find one that matches the value you selected. If you find a match, the package is currently installed; if you do not find a match, the package has been lost and you must reinstall it using the MTS administrator user interface.

5. Obtain the **GUID** value (a unique hexadecimal number string) of the component in question. Check the list of subkeys under the package until you find one that matches the value you selected. If you find a match, the component is currently installed in that package; if you do not find a match, the package has been lost and you must reinstall it using the MTS administrator user interface.

TIP: You can obtain the GUID value of a package from the MTS administrator user interface by selecting a package and displaying its Properties dialog. The General tab identifies a Package ID, which is the GUID you need.

Determining the Roles Permitted to Access an MTS Package

Another Registry setting determines which MTS security roles are allowed to access a given MTS package. To access this Registry setting, follow these steps:

1. Launch Regedt32.

2. Select the Window menu option for **HKEY_LOCAL_MACHINE**.

3. Use the tree control in the left-hand window to navigate to the **SOFTWARE\Microsoft\Transaction Server\Packages** subkey. Double-click on the subkey to expand it and display its values in the right-hand window.

4. Obtain the **GUID** value (a unique hexadecimal number string) of the package in question. Check the list of subkeys for packages until you find one that matches the value you selected. If you find a match, the package is currently installed; if you do not find a match, the package has been lost and you must reinstall it using the MTS administrator user interface. Repeat this process on each of the selected package's subkeys, checking for the role's GUID value (obtained via the Properties dialog box for the role in the administrator user interface as the Role ID). If you don't find this GUID entry, you must use the MTS Explorer to re-create the role entry for that component.

5. If you locate the GUID for the role in the package's subkeys, the role is allowed to access the package; otherwise, the role is not allowed to access the package, and you must enable it using the MTS administrator user interface.

TIP: *You can obtain the GUID value of a package from the MTS administrator user interface by selecting a package and displaying its Properties dialog box. The General tab identifies a Package ID, which is the GUID you need.*

Determining the Roles Permitted to Access an MTS Component

Another Registry setting determines which MTS security roles are allowed to access an MTS component inside a package. To access this Registry setting, follow these steps:

1. Launch Regedt32.

2. Select the Window menu option for **HKEY_LOCAL_MACHINE**.

3. Use the tree control in the left-hand window to navigate to the **SOFTWARE\Microsoft\Transaction Server\Packages** subkey. Double-click on the subkey to expand it and display its values in the right-hand window.

4. Obtain the **GUID** value (a unique hexadecimal number string) of the package in question. Check the list of subkeys for packages until you find one that matches the value you selected. If you find a match, the package is currently installed; if you do not find a match, the package has been lost and you must reinstall it using the MTS administrator user interface. Repeat this process on each of the selected package's subkeys, checking for the role's GUID value (obtained via the Properties dialog box for the role in the administrator user interface as the Role ID).

5. If you locate the GUID for a role in the component's subkeys, the role is allowed to access that component's interfaces; otherwise, the role is not allowed to access the component's interfaces, and it must be enabled using the MTS administrator user interface.

TIP: *You can obtain the GUID value of a package from the MTS administrator user interface by selecting a package and displaying its Properties dialog box. The General tab identifies a Package ID, which is the GUID you need.*

Determining the Computers Added to an MTS System

Another Registry setting lists all the computers currently installed for use by MTS. To view which computers MTS can use, follow these steps:

1. Launch Regedt32.

2. Select the Window menu option for **HKEY_LOCAL_MACHINE**.

3. Use the tree control in the left-hand window to navigate to the **SOFTWARE\Microsoft\Transaction Server\Computers** subkey. Double-click on the subkey to expand it and display its values in the right-hand window.

4. All the computers that MTS currently supports are shown as subkeys with their names. If a particular computer is not found (or its name is incorrect), you can reinstall it by using the MTS administrator user interface.

Determining if a Component Has Been Installed Remotely on an MTS System

Another Registry setting indicates whether a component is installed for remote access by MTS (that is, entries have been placed in the Registry so that MTS can determine which remote computer to use to create an instance of the component). To access this Registry setting, follow these steps:

1. Launch Regedt32.

2. Select the Window menu option for **HKEY_LOCAL_MACHINE**.

3. Use the tree control in the left-hand window to navigate to the **SOFTWARE\Microsoft\Transaction Server\Remote Components** subkey. Double-click on the subkey to expand it and display its values in the right-hand window.

4. Obtain the **GUID** value (a unique hexadecimal number string) of the component in question. Check the list of subkeys for components until you find one that matches the value you selected. If you find a match, the component is currently installed; if you do not find a match, the component has been lost and you must reinstall it using the MTS administrator user interface.

Chapter 20

SQL Server

In Brief

SQL Server is Microsoft's competitor in the relational database management system (RDBMS) market, going against products such as Oracle and Informix. SQL Server runs on Windows NT/2000 Server as well as a number of other Microsoft operating systems. It provides support for Transact-SQL, a dialect of SQL (Structured Query Language), the industry standard for interaction with RDBMS applications. SQL Server features distributed transactions, database replication, and a number of other powerful features. SQL Server is supported by the ADO (Active Data Object) and RDO (Remote Data Object) database access technologies from Microsoft, and supports the Open Database Connectivity (ODBC) standard.

SQL Server Features

Besides including its own feature set, SQL Server allows the use of several other powerful Microsoft products and technologies:

- Distributed transactions (transactions spanning two different databases) with the Distributed Transaction Coordinator (DTC)

- Database replication

- Web server connections

- Windows 2000 event logging

- Multiple networking protocols

- Support for ODBC

- Heterogeneous transactions (different database systems) using Microsoft Transaction Server (MTS)

ODBC

ODBC is an industry standard that permits any vendor to develop applications independent of the internal structure of the implementing database. SQL Server fully supports ODBC, allowing it to work with RDBMS applications and components from many different vendors.

ADO and RDO

Active Data Objects and Remote Data Objects are the two major database access technologies currently used by Microsoft. Due to its support for ODBC, SQL Server is fully accessible to both ADO and RDO components and applications.

SQL Server Services

SQL Server consists of several services that provide special capabilities to the system, which are normally accessed via the Enterprise Manager (shown in Figure 20.1). SQL Server Enterprise Manager provides access to the SQL Executive, SQL-DMO, and SQL Replication services. Enterprise Manager is itself a Microsoft Management Console (MMC) application and can be merged or augmented with other MMC applications, such as Component Services.

SQL Executive

SQL Executive is the service application that manages connections, access, and logging of SQL Server events. It must be running for any other SQL Server functionality to be available.

SQL-DMO

SQL-DMO (SQL Distributed Management Objects) is SQL Server enterprise-level capability (actually, a programming interface) that

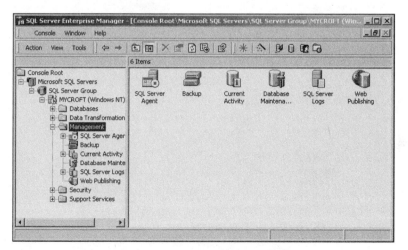

Figure 20.1   The SQL Server Enterprise Manager user interface in Windows 2000.

allows remote components to use a central SQL Server database. SQL-DMO requires several additional installations on both the host and client machines to function.

SQL Replication

SQL Server can have its database replicated for backup purposes and for the distribution of data throughout the organization. The SQL Executive controls the SQL Replication feature. Replication is a complex process involving a scheduling engine, task lists, synchronization, and subscribers.

Immediate Solutions

Configuring the Correct Named Pipe for SQL Server's Network Library

SQL Server Executive needs a trusted connection, which in practice usually means using Named Pipes. A problem can arise when parallel installations of SQL Server are placed on the same machine—the necessary Named Pipe can be changed or lost. Here's how to change the Registry to fix this problem:

1. Launch Regedt32.

2. Select the Window menu option for **HKEY_LOCAL_MACHINE**.

3. Use the tree control in the left-hand window to navigate to the **SOFTWARE\Microsoft\MSSQLServer\SQL Executive** subkey. Click on the subkey to select it and display its values in the right-hand window.

4. Locate the **ServerHost** value entry. Use the String Editor to change the ServerHost value to the desired Named Pipe.

Fixing Slow SQL Server Named Pipe Connections with Windows 9x Workstations

SQL Server uses Named Pipes heavily and this can result in very slow connections with Windows 9x workstations due to a specific Registry setting. Here's how to change the Registry to correct the problem:

1. On the Windows 9x machine, run Regedit.exe.

2. Select the Window menu option for **HKEY_LOCAL_MACHINE**.

3. Use the tree control in the left-hand window to navigate to the **SYSTEM\CurrentControlSet\Services\VxD\VNETSUP** subkey. Click on the subkey to select it and display its values in the right-hand window.

4. Create a new value named **DirectHost**, of type REG_SZ. Set DirectHost to 0. This turns off Direct Hosting for that workstation and fixes the slow connection problem.

Increasing Available DB-Library Connections for SQL Server

SQL Server has a hard-wired number of DB-Library concurrent connections for SQL Executive. If a system can support a higher number, administrators can increase this setting via the following Registry tweak:

1. Launch Regedt32.

2. Select the Window menu option for **HKEY_LOCAL_MACHINE**.

3. Use the tree control in the left-hand window to navigate to the **SOFTWARE\Microsoft\MSSQLServer\SQL Executive** subkey. Click on the subkey to select it and display its values in the right-hand window.

4. Add a new value entry named **MaxDBProcesses** of type REG_DWORD. Set MaxDBProcesses to the number of desired concurrent DB-Library connections. The actual number available is 3 fewer due to system overhead. This technique works only for SQL Server 6.5 and higher.

TIP: The default for MaxDBProcesses is 61.

Preventing SQL Server Startup Failure Due to Slow RPC Initialization

A rare but annoying error can occur when a server's RPC (remote procedure call) service is slow to initialize and SQL Server (configured to use Multiprotocol) starts before the RPC service finishes. This causes SQL Server to fail because SQL depends on the RPC service. Fortunately, there is a Registry change that prevents this problem:

1. Launch Regedt32.

2. Select the Window menu option for **HKEY_LOCAL_MACHINE**.

3. Use the tree control in the left-hand window to navigate to the **SYSTEM\CurrentControlSet\Services\MSSQLServer** subkey. Click on the subkey to select it and display its values in the right-hand window.

4. Create a new value entry named **DependOnService** with a type of REG_MULTI_SZ. Enter the string "RpcSs" for its setting. Restart the server and the problem should be corrected.

Fixing SQL Server Executive Service Logon Failures

SQL Server sometimes fails to start properly because the Executive Service cannot log on. One of the principal causes of this is that the system fails to grant full access to two Registry keys to the Local System Account and Administrator Group. To access the two keys, follow these steps:

1. Launch Regedt32.

2. Select the Window menu option for **HKEY_LOCAL_MACHINE**.

3. Use the tree control in the left-hand window to navigate to the **SYSTEM\CurrentControlSet\Services\SQLExecutive** subkey. Click on the subkey to select it and display its values in the right-hand window.

4. Ensure that both the Administrator Group and Local System Account have full control privileges for this key and subkeys.

5. Repeat Steps 3 and 4 for the **SYSTEM\CurrentControlSet\ Services\MSSQLServer\SQLExecutive** subkey.

Fixing SQL Server Total Task Failures

SQL Server can have several levels of failure, including partial task execution failures and complete task execution failures. In the latter case, one often-overlooked problem is a conflicting Registry entry for the use of Named Pipes. To correct this conflict, follow these steps:

1. Launch Regedt32.

2. Select the Window menu option for **HKEY_LOCAL_MACHINE**.

3. Use the tree control in the left-hand window to navigate to the **SOFTWARE\Microsoft\MSSQLServer\SQL Executive** subkey. Click on the subkey to select it and display its values in the right-hand window.

4. Locate the **ServerHost** value entry. Use the String Editor to change its value so that it does not include the server name (only the reference to the desired Named Pipe).

Configuring SQL Server Executive to Write a Verbose Log File

Sometimes debugging the SQL Executive becomes necessary. In this case, you can use the following Registry tweak to cause SQL Executive to write a verbose log file:

1. Launch Regedt32.

2. Select the Window menu option for **HKEY_LOCAL_MACHINE**.

3. Use the tree control in the left-hand window to navigate to the **SOFTWARE\Microsoft\MSSQLServer\SQL Executive** subkey. Click on the subkey to select it and display its values in the right-hand window.

4. Add a value entry named **VerboseFile**, of data type REG_SZ. Set VerboseFile to the full path and file name for the log file desired. Stop and restart SQL Executive and it writes a verbose log to this location until you remove the VerboseFile entry.

Fixing SQL Server Executive Version Detection Failures

A problem that sometimes happens with SQL Server's Executive service is that the service fails to start due to an inability to detect the current SQL Server version number. This is caused by a Registry permission problem that you can fix by giving the Administrator Group and Local System Account full privileges in the Current Version Registry key as follows:

1. Launch Regedt32.

2. Select the Window menu option for **HKEY_LOCAL_MACHINE**.

3. Use the tree control in the left-hand window to navigate to the **SYSTEM\CurrentControlSet\Services\SQLExecutive\Current Version** subkey. Click on the subkey to select it and display its values in the right-hand window.

4. Make sure that both the Administrator Group and Local System Account have full control privileges for this key and its subkeys.

Configuring SQL Server Asynchronous Query Processing Timeouts

There is a built-in sleep period for all SQL Server DB-Library Asynchronous Query processing requests. In some cases, this can negatively impact performance. Administrators can change this sleep period via the following Registry tweak:

1. Launch Regedt32.
2. Select the Window menu option for **HKEY_LOCAL_MACHINE**.
3. Use the tree control in the left-hand window to navigate to the **SOFTWARE\Microsoft\MSSQLServer\Client\DB-Lib** subkey. Click on the subkey to select it and display its values in the right-hand window.
4. Locate the **DataReadySleep** value entry. Use the DWORD Editor to set DataReadySleep's value to a lower one if needed.

TIP: *DataReadySleep's range is from 0 through 1,000 milliseconds, with a default of 250. Setting the DataReadySleep value to **0xFFFFFFFF** disables the sleep interval entirely.*

WARNING! Setting the DataReadySleep value to 0 may prevent asynchronous query processing from ever executing!

Preventing SQL Server Web Server Connection Resets

If SQL Server is simultaneously requested over TCP/IP by several clients, it sometimes resets the connections. One possible way of eliminating this problem is to change the Registry as follows:

1. Launch Regedt32.
2. Select the Window menu option for **HKEY_LOCAL_MACHINE**.
3. Use the tree control in the left-hand window to navigate to the **SOFTWARE\Microsoft\MSSQLServer\MSSQLServer** subkey. Click on the subkey to select it and display its values in the right-hand window.
4. Locate the **WinsockListenBacklog** value entry. Use the DWORD Editor to set WinsockListenBacklog's value to a higher value until the problem stops.

20. SQL Server

> **TIP:** *WinSockListenBacklog's range is from **1** through **0xFFFFFFFF**, with a default of 100.*

> **WARNING!** *Due to the nature of the reset problem, administrators might have to use trial and error to find the right setting for the WinSockListenBacklog's entry to eliminate the problem!*

Configuring SQL Server Replication

In rare cases, SQL Server replication can fail due to invalid or corrupted Registry entries for its distribution database and path. To check for such a problem and (possibly) repair it, the administrator should check the Registry as follows:

1. Launch Regedt32.

2. Select the Window menu option for **HKEY_LOCAL_MACHINE**.

3. Use the tree control in the left-hand window to navigate to the **SOFTWARE\Microsoft\MSSQLServer\Replication** subkey. Click on the subkey to select it and display its values in the right-hand window.

4. Locate the **DistributionDB** and **WorkingDirectory** value entries. Use the String Editor to make sure DistributionDB and WorkingDirectory are set to the proper values.

> **WARNING!** *Only edit DistributionDB and WorkingDirectory manually in failure situations to avoid introducing unpredictable behavior.*

Chapter 21

Microsoft Office

In Brief

The Microsoft Office suite enjoys incredible popularity, so this is one chapter you should not skip. Microsoft Office includes Word, PowerPoint, Access, Outlook, and Excel— predominant applications in today's business world. This chapter covers a number of important Registry locations and modifications for the Office suite.

Microsoft Office

Microsoft Office is the best-selling office suite, bar none. It has almost become the standard for general business applications and any Windows 2000 administrator had better be ready to support this group of applications.

Access

Microsoft Access stores an incredible amount of data for small to mid-size companies. Access makes it very easy to ask for the information managers need, and provides powerful tools that help organize and share database content. The latest versions of Microsoft Access also make it very easy to share this information over corporate intranets or the public Internet.

Excel

Microsoft Excel permits users to take advantage of comprehensive tools to create spreadsheets and share them on the Web for universal viewing and collaboration. The application also enables data analysis using charts, Microsoft PivotTable views, and graphs. In addition, automatic formatting helps create data-rich spreadsheets.

PowerPoint

Microsoft PowerPoint provides a complete set of tools for creating powerful presentations. Users can organize and format their material easily, illustrate points with clip art, and even broadcast presentations over the Web.

Word

Microsoft Word is a word processor and much more, thanks to many enhanced desktop publishing features. Like its Office suite siblings, recent versions of the application stress the creation of Web-enabled documents for easy collaboration via the Internet or corporate intranet.

Outlook

Microsoft Outlook provides a single location for organizing and managing all day-to-day information, from email and calendars, to contacts and task lists. The email client integrates seamlessly with many client/server messaging platforms and protocols.

Immediate Solutions

Checking the Microsoft Office Installation Path

Many applications that are designed to work with Microsoft Office check the Registry for the installation path of the Office suite. This path is found in the following Registry location:

1. Launch Regedt32.
2. Select the Window menu option for **HKEY_LOCAL_MACHINE**.
3. Use the tree control in the left-hand window to navigate to the **SOFTWARE\Microsoft\Office*[version number]*\\Common\InstallRoot** subkey, where *[version number]* is your version of Microsoft Office. Click on the subkey to select it and display its values in the right-hand window.
4. Locate to the **Path** value. This Reg_SZ value is the installation path.

Change Help Links to Office Web Sites

You might have noticed that later versions of Office feature Help menus and alerts that offer to take the user to the Microsoft Office Web site. You can change the behavior of these menu items and alerts using the following Registry tweaks:

1. Launch Regedt32.
2. Select the Window menu option for **HKEY_CURRENT_USER**.
3. Use the tree control in the left-hand window to navigate to the **SOFTWARE\Microsoft\Office*[version number]*\\Common\General** subkey, where *[version number]* is your version of Microsoft Office. Click on the subkey to select it and display its values in the right-hand window.
4. Locate the **CustomizableAlertBasedURL** value and double-click it to open it in the string editor. Change the location string value—or delete it if you want to prevent users from visiting

the Web with this feature. To do the same for Microsoft Office
alerts that offer to take the user to the Web, change the
CustomizableAlertDefaultButtonText as above.

Deleting a File from the Most Recently Used List in Access

When you launch Microsoft Access, you are presented with the option of opening a particular database that you have used recently. To remove an entry from this list, follow these steps:

1. Launch Regedt32.

2. Select the Window menu option for **HKEY_CURRENT_USER**.

3. Use the tree control in the left-hand window to navigate to the
   **SOFTWARE\Microsoft\Office*[version number]*\\
   Access\Settings** subkey, where *[version number]* is your
   version of Microsoft Office. Click on the subkey to select it and
   display its values in the right-hand window.

4. Double click the **MRU** value that contains the entry you want to
   remove. Delete the entry in the String Editor and choose OK.

Deleting a File from the Most Recently Used List in Excel

The File menu in Excel displays a list of files that you have opened
recently. To delete a file from this list, follow these instructions:

1. Launch Regedt32.

2. Select the Window menu option for **HKEY_CURRENT_USER**.

3. Use the tree control in the left-hand window to navigate to the
   **SOFTWARE\Microsoft\Office*[version number]*\Excel\\
   Recent Files** subkey, where *[version number]* is your
   version of Microsoft Office. Click on the subkey to select it and
   display its values in the right-hand window.

4. Delete the appropriate **FileX** value where X is a number
   representing the file you want to delete from the list.

Deleting a File from the Most Recently Used List in PowerPoint

The File menu in PowerPoint displays a list of files that you have opened recently. To delete a file from this list, follow these instructions:

1. Launch Regedt32.

2. Select the Window menu option for **HKEY_CURRENT_USER**.

3. Use the tree control in the left-hand window to navigate to the **SOFTWARE\Microsoft\Office*version number*** **PowerPoint\Recent File List** subkey, where *[version number]* is your version of Microsoft Office. Click on the subkey to select it and display its values in the right-hand window.

4. Delete the appropriate **FileX** value where X is a number representing the file you want to delete from the list.

Deleting a File from the Most Recently Used List in Word

Actually, deleting a file from the most recently used list in Word does not directly involve the Registry, but we thought we would include it here for consistency's sake with the other Immediate Solutions in this chapter. To delete a particular file, follow these instructions:

1. While in Word, press Ctrl+Alt+- (hyphen) on the keyboard.

2. Select the File menu and click on the file that you want removed from the list.

Deleting a Most Recently Used Typeface in PowerPoint

PowerPoint tracks the typefaces used by authors and displays these at the top of the Font dropdown list on the Formatting toolbar. If you want to delete specific typefaces from this list, follow these steps:

1. Launch Regedt32.

2. Select the Window menu option for **HKEY_CURRENT_USER**.

3. Use the tree control in the left-hand window to navigate to the **SOFTWARE\Microsoft\Office*version number*\\ PowerPoint\Recent Typeface List** subkey, where *version number]* is your version of Microsoft Office. Click on the subkey to select it and display its values in the right-hand window.

4. Delete the appropriate **TypefaceX** value, where X is the number associated with the typeface you want to delete.

Checking the Access Install Location

If you need a quick way to check where Microsoft Access was installed or if you are having difficulty installing applications that function with Access, follow these steps to check this information in the Registry:

1. Launch Regedt32.

2. Select the Window menu option for **HKEY_LOCAL_MACHINE**.

3. Use the tree control in the left-hand window to navigate to the **SOFTWARE\Microsoft\Office*version number*\Access\ InstallRoot** subkey, where *version number]* is your version of Microsoft Office. Click on the subkey to select it and display its values in the right-hand window.

4. The **Path** value displays the installation path for Access.

Checking the Workgroup Information File Location for Access

While Access is not known for its security mechanisms, it does allow you to secure databases. A Microsoft Jet database stores the definitions of users, groups, and passwords that an Access administrator uses in the security system. Where is this important database stored? The Registry contains the answer:

1. Launch Regedt32.

2. Select the Window menu option for **HKEY_LOCAL_MACHINE**.

3. Use the tree control in the left-hand window to navigate to the **SOFTWARE\Microsoft\Office\\[*version number*]\Access\ Jet\\[Jet version number]\Engines** subkey, where *[version number]* is your version of Microsoft Office and *[Jet version number]* is your version of Microsoft Jet. Click on the subkey to select it and display its values in the right-hand window.

4. The **SystemDB** value contains the path to the Workgroup Information File controlling security in Access.

Checking the Excel Install Location

If you need a quick way to check where Microsoft Excel is installed, or if you are having difficulty installing applications that function with Excel, follow these steps to check this information in the Registry:

1. Launch Regedt32.

2. Select the Window menu option for **HKEY_LOCAL_MACHINE**.

3. Use the tree control in the left-hand window to navigate to the **SOFTWARE\Microsoft\Office\\[*version number*]\Excel\ InstallRoot** subkey, where *[version number]* is your version of Microsoft Office. Click on the subkey to select it and display its values in the right-hand window.

4. The **Path** value displays the installation path for Access.

Checking the Word Install Location

If you need a quick way to check where Microsoft Word was installed, or if you are having difficulty installing applications that function with Word, follow these steps to check this information in the Registry:

1. Launch Regedt32.

2. Select the Window menu option for **HKEY_LOCAL_MACHINE**.

3. Use the tree control in the left-hand window to navigate to the **SOFTWARE\Microsoft\Office\\[*version number*]\Word\ InstallRoot** subkey, where *[version number]* is your version of Microsoft Office. Click on the subkey to select it and display its values in the right-hand window.

4. The **Path** value displays the installation path for Access.

Resetting User Options in Word

If you have changed configurations in Word and you want to return to the default set of configuration options, follow these steps:

1. Launch Regedt32.
2. Select the Window menu option for **HKEY_CURRENT_USER**.
3. Use the tree control in the left-hand window to navigate to the **SOFTWARE\Microsoft\Office\[*version number*]\ Word\Data** subkey, where *[version number]* is your version of Microsoft Office. Click on the subkey to select it and display its values in the right-hand window.
4. Delete the **Data** subkey.

Changing the Default WordMail Template

Word has many very convenient features. One of these is WordMail, which allows Word to compose and send e-mail messages. You can control the default template for WordMail by editing a Registry value as follows:

1. Launch Regedt32.
2. Select the Window menu option for **HKEY_CURRENT_USER**.
3. Use the tree control in the left-hand window to navigate to the **SOFTWARE\Microsoft\Office\[*version number*]\Word\ Stationary** subkey, where *[version number]* is your version of Microsoft Office. Click on the subkey to select it and display its values in the right-hand window.
4. The **Default Template** setting contains the path to the default template. This is email.dot by default. To change it, use the String Editor.

Checking the Outlook Install Location

If you need a quick way to check where Microsoft Outlook was installed, or if you are having difficulty installing applications that function with Outlook, follow these steps to check this information in the Registry:

1. Launch Regedt32.
2. Select the Window menu option for **HKEY_LOCAL_MACHINE**.

3. Use the tree control in the left-hand window to navigate to the **SOFTWARE\Microsoft\Office\\[version number]\Outlook\ InstallRoot** subkey, where **[version number]** is your version of Microsoft Office. Click on the subkey to select it and display its values in the right-hand window.

4. The **Path** value displays the installation path for Access.

Returning Microsoft Outlook to the First-Run Condition

If you need to return Microsoft Outlook to its first-run state on a system, follow these instructions:

1. Launch Regedt32.

2. Select the Window menu option for **HKEY_CURRENT_USER**.

3. Use the tree control in the left-hand window to navigate to the **SOFTWARE\Microsoft\Office\\[version number]\Outlook** subkey, where **[version number]** is your version of Microsoft Office. Click on the subkey to select it and display its values in the right-hand window.

4. Select the **FirstRunDialog** value and use the String Editor to change the value to True.

5. If you also want to re-create all sample welcome items, navigate to the **SOFTWARE\Microsoft\Office\\[version number]\ Outlook\Setup** subkey, where **[version number]** is your version of Microsoft Office. Click on the subkey to select it and display its values in the right-hand window.

6. Select and delete the **CreateWelcome** and **First-Run** values.

TIP: *You can delete the mail account settings on the system by right-clicking the Outlook icon on your Desktop and choosing Properties. Select Show Profiles in order to access the dialog for deleting mail accounts.*

Fixing Outlook Journaling

If you have configured Outlook to use the journaling feature, but the journaling data is not stored in the Journal folder, it is most likely because a specific Registry value does not point to a valid folder location in the operating system file structure. To fix this, follow these steps:

1. Launch Regedt32.

2. Select the Window menu option for **HKEY_CURRENT_USER**.

3. Use the tree control in the left-hand window to navigate to the **SOFTWARE\Microsoft\Office\[version number]\ Outlook\Journal** subkey, where *[version number]* is your version of Microsoft Office. Click on the subkey to select it and display its values in the right-hand window.

4. Ensure the **Item Log File** value points to a valid location. If it does not, use the String Editor to modify the value.

Hiding the Outlook Public Folders

Determining whether the public folders are visible or not is controlled with a simple Registry setting as follows:

1. Launch Regedt32.

2. Select the Window menu option for **HKEY_LOCAL_MACHINE**.

3. Use the tree control in the left-hand window to navigate to the **SOFTWARE\Microsoft\Office\[version number]\ Outlook\Dataviz** subkey, where *[version number]* is your version of Microsoft Office. Click on the subkey to select it and display its values in the right-hand window.

4. The **HidePublicFolders** value defaults to 1. In order to display the folders—change this value to 0.

Checking the PowerPoint Install Location

If you need a quick way to check where Microsoft PowerPoint was installed, or if you are having difficulty installing applications that function with PowerPoint, follow these steps to check this information in the Registry:

1. Launch Regedt32.

2. Select the Window menu option for **HKEY_LOCAL_MACHINE**.

3. Use the tree control in the left-hand window to navigate to the **SOFTWARE\Microsoft\Office\\[*version number*]\\ PowerPoint\InstallRoot** subkey, where *[version number]* is your version of Microsoft Office. Click on the subkey to select it and display its values in the right-hand window.

4. The **Path** value displays the installation path for Access.

Fixing Microsoft Clip Gallery

The Microsoft Clip Gallery brings clip art to the Office applications in a very convenient and integrated manner. Unfortunately, Clip Gallery can fall prey to a number of errors in Windows 2000. Here is a list of the errors you may experience:

- Server Application or Source File cannot be found

- Invalid Page Fault in module <Name of Module>

- Clip Gallery cannot run because the database has been marked read-only

- Clip Gallery does not start and no errors are displayed

A key troubleshooting step for all these errors is to check the Registry entries that control Microsoft Clip Gallery. Follow these steps:

1. Launch Regedt32.

2. Select the Window menu option for **HKEY_CLASSES_ROOT**.

3. Use the tree control in the left-hand window to navigate to the **MS_ClipArt_Gallery\CLSID** subkey. Click on the subkey to select it and display its values in the right-hand window.

4. Check the default value—It should be a value in the following format: **{00030026-0000-0000-C000-000000000046}**.

5. Use the tree control in the left-hand window to navigate to the **MS_ClipArt_Gallery\protocol\StdFileEditing\server** subkey. Click on the subkey to select it and display its values in the right-hand window.

6. Ensure the default value contains the following string: **C:\PROGRA~1\COMMON~1\MICROS~1\Artgalry\artgalrY.exe**.

7. Use the tree control in the left-hand window to navigate to the **MS_ClipArt_Gallery.2\CLSID** subkey. Click on the subkey to select it and display its values in the right-hand window.

8. Check the default value—it should be a value in the following format: **{00021290-0000-0000-C000-000000000046}**.

9. Use the tree control in the left-hand window to navigate to the **MS_ClipArt_Gallery.2\protocol\StdFileEditing\server** subkey. Click on the subkey to select it and display its values in the right-hand window.

10. Ensure the default value contains the following string: **C:\PROGRA~1\COMMON~1\MICROS~1\Artgalry\artgalrY.exe**.

Index

A

Access. *See* Microsoft Access.
AccessDeniedMessage value, 302
Access level
   parallel port, 138
   serial port, 139
Access tokens, 225
Acknowledgment piggybacking,
   175–176
AckWindow value, 175
Active Data Objects. *See* ADO.
Active Desktop, 316, 325, 330–334
Active Directory, 111
   Group Policy snap-in, 34–35, 49
   restoring, 55
ActiveMovie, 256
ActiveWindowTracking value, 105
AddNameQueryTimeout value, 188
Address bar, IE4+, 324
Addressing, NDIS IEEE address,
   212–213
Address Resolution Protocol.
   See ARP.
Address resolution timeouts, 187–188
Administration tools. *See* System
   administration tools.
Administrative Alert, 40, 156
Administrative Templates section,
   Group Policy, 34
ADO, 357
Alerter service, 155, 156
AlertNames value, 156
AllocateCDRoms value, 231
allocatedasd value, 232
AllocateFloppies value, 231
AllowAnonymous value, 296
AllowGuestAccess value, 303, 308
AllowKeepAlives value, 308

Animation, disabling "extra"
   Windows animations, 94
Anonymous FTP, 288
Anonymous HTTP, 288
Anonymous logins, IIS, 296
AnonymousUserName value, 296
Anonymous users, IIS, 295–296
APIProtocolSupport value, 118
AppleTalk, 174, 179, 183
AppleTalk router, 183
ARP, RRAS, 208
ArpUseEtherSNAP value, 112
AsyncMac frame sizes, 212
Attrib command, 55
Audio services, 257
Auditing, 207, 209, 226
AuthenticateRetries value, 210
AuthenticateTime value, 210
Authentication, RRAS, 210
Authorization level, COM+, 281
AutoAdminLogon value, 232
AutoDisconnect value, 211
AutoEndTasks value, 151
Autoexec.bat files, **PATH**
   statement, 150
Automatic reboot, 96
Auto Refresh option, Regedt32, 15
AutoRestartShell value, 96
Autorun value, 103

B

Background color, customizing, 100
Background sound, IE4+, 320
Background value, 100
BackupDatabasePath value, 119
Backups, 56–57, 58, 63–64
Batch files, automatic Registry
   updates, 33

L

Build your IT credentials with these titles from Certification Insider Press!

**MCSE Windows 2000
Directory Services Design
Exam Cram**
ISBN: 1-57610-714-0

**MCSE Windows 2000
Network Design
Exam Cram**
ISBN: 1-57610-716-7

**MCSE Windows 2000
Security Design
Exam Cram**
ISBN: 1-57610-715-9

**MCSE Windows 2000
Foundations
Exam Prep**
ISBN: 1-57610-679-9

Expand your Windows 2000 knowledge with these reference guides from Coriolis Technology Press!

**Windows 2000
Security
Little Black Book**
ISBN: 1-57610-387-0

**Windows 2000
Reducing TCO
Little Black Book**
ISBN: 1-57610-315-3

**Windows 2000 Registry
Little Black Book,
2nd Edition**
ISBN: 1-57610-882-1

**Windows 2000 Server
Architecture and Planning,
2nd Edition**
ISBN: 1-57610-607-1

The Coriolis Group, LLC

Telephone: 480.483.0192 • Toll-free: 800.410.0192 • www.coriolis.com

Coriolis books are also available at bookstores and computer stores nationwide.